AF573885

Lecture Notes in Electrical Engineering

Volume 557

The book series *Lecture Notes in Electrical Engineering* (LNEE) publishes the latest developments in Electrical Engineering - quickly, informally and in high quality. While original research reported in proceedings and monographs has traditionally formed the core of LNEE, we also encourage authors to submit books devoted to supporting student education and professional training in the various fields and applications areas of electrical engineering. The series cover classical and emerging topics concerning:

- Communication Engineering, Information Theory and Networks
- Electronics Engineering and Microelectronics
- Signal, Image and Speech Processing
- Wireless and Mobile Communication
- Circuits and Systems
- Energy Systems, Power Electronics and Electrical Machines
- Electro-optical Engineering
- Instrumentation Engineering
- Avionics Engineering
- Control Systems
- Internet-of-Things and Cybersecurity
- Biomedical Devices, MEMS and NEMS

For general information about this book series, comments or suggestions, please contact leontina.dicecco@springer.com.

To submit a proposal or request further information, please contact the Publishing Editor in your country:

**** Indexing: The books of this series are submitted to ISI Proceedings, EI-Compendex, SCOPUS, MetaPress, Web of Science and Springerlink ****

More information about this series at http://www.springer.com/series/7818

Tohid Jahangiri · Qian Wang ·
Filipe Faria da Silva · Claus Leth Bak

Electrical Design of a 400 kV Composite Tower

Tohid Jahangiri
Department of Energy Technology
Aalborg University
Aalborg East, Denmark

Qian Wang
Department of Energy Technology
Aalborg University
Aalborg East, Denmark

Filipe Faria da Silva
Department of Energy Technology
Aalborg University
Aalborg East, Denmark

Claus Leth Bak
Department of Energy Technology
Aalborg University
Aalborg East, Denmark

ISSN 1876-1100 ISSN 1876-1119 (electronic)
Lecture Notes in Electrical Engineering
ISBN 978-3-030-17842-0 ISBN 978-3-030-17843-7 (eBook)
https://doi.org/10.1007/978-3-030-17843-7

This Springer imprint is published by the registered company Springer Nature Switzerland AG
The registered company address is: Gewerbestrasse 11, 6330 Cham, Switzerland

Preface

The global goals of decarbonizing energy supply call for alternative sources of generation of electric power. Such are in nature located geographically elsewhere than the conventional power plants. Offshore wind power plants are located in the ocean. Solar plants are located in areas with a high intensity of solar radiation. Altogether, the replacement of conventional power plants often located close to centers of consumption with more remote sources creates a huge demand for restructuring the transmission grid. New lines become necessary to transport the renewable energy. Overhead lines are the dominant and only realistic way to transport large quantities of electric power over long distances.

Overhead line technology reaches back a century without much change in design. However, nowadays the public opinion against overhead lines is very negative and new lines often are delayed by tens of years due to governmental processing to get public acceptance. An entirely new, more modern and visually friendly look of overhead lines would undoubtedly ease this process leading to a faster reaching of the climate goals.

The Power Pylons of the Future (PoPyFu) project aims at introducing a groundbreaking new design for transmission overhead line design based on a fully composite tower with a less visible and more visually friendly design, which is also expected to accelerate the construction of new transmission corridors at a lower cost.

The authors of this book are Dr. Tohid Jahangiri, Dr. Qian Wang, Prof. Claus Leth Bak and Associate Professor Filipe Faria da Silva from Department of Energy Technology, Aalborg University, Denmark. This book is a part of the Power Pylons of the Future (PoPyFu) project, which has been funded by the Danish Research Council (Innovationsfonden), to whom we would like to thank the financial support of this project. This book provides researchers, developers and practitioners with a major contribution to the design and evaluation of the initial concept of a fully composite pylon, based on theoretical studies, finite element method and experimental results.

We would like to express our gratitude to:
Prof. Joachim Holbøll, Technical University of Denmark, Denmark.
Mr. Henrik Skouboe, Bystrup Company, Denmark.
Prof. Christian Franck, Swiss Federal Institute of Technology (ETH), Switzerland.
All project partners of the Power Pylons of the Future (PoPyFu) project.

Aalborg, Denmark
February 2019

Tohid Jahangiri
Qian Wang
Claus Leth Bak
Filipe Faria da Silva

Contents

Chapter 1
Overview of Composite-Based Transmission Pylons

1.1 Introduction

Overhead transmission lines are one of the most important parts of high voltage (HV) power systems and are essential to connect power plants to the load centers. The transition from non-renewable energy sources to renewable energy sources such as wind, hydro and solar powers leads to the increasing demand for the planning of new overhead transmission lines. Regardless of overhead transmission lines, underground power cables can be used to transport electric power at high voltage (HV) level. This alternative is 6–12 times more expensive than overhead line and is technically suitable for short distances [1]. Depending on the distance of the line, the transmission of electric power by overhead lines is more economical at extra-high voltage (EHV) and ultra-high voltage (UHV) levels due to lower losses along the lines at EHV and UHV levels.

However, most of today's overhead lines are based on conventional steel lattice towers which have been developed and utilized since 70 years ago. Other overhead line towers or poles based on wood, concrete, and composite materials are also available but their application is limited to HV levels or below [2].

In last decades, public opposition has increased against existing overhead lines (with steel lattice towers) and installation of new overhead lines. The main reasons for the opposition are the proximity of overhead lines with their accommodations, which arises a fear for the possible effects of low-frequency electromagnetic emissions on human health near the lines, and the visual impact of overhead lines on environment. The last case is due to this fact that steel lattice towers are very dominant in landscapes and have an unpleasing aesthetic view. Therefore, there is a need for modern designs of power pylons which satisfy general public opinion and at the same time reduce the electromagnetic emissions at the right-of-way (ROW) of overhead lines.

T. Jahangiri et al., *Electrical Design of a 400 kV Composite Tower*, Lecture Notes in Electrical Engineering 557,
https://doi.org/10.1007/978-3-030-17843-7_1

1.2 Composite-Based Transmission Towers-State of the Art Review

1.2.1 Demand for New Overhead Lines

The world is in an ever-increasing demand of electricity power [3]. Meanwhile, it is faced with the fact of limited non-renewable energy sources and risk of environmental destruction due to over development of non-renewable sources [3]. In order to fulfill the requirement for increasing amount of electricity power per year without ruin the environment, development of renewable sources, such as wind, solar, tides, waves and geothermal heat, is necessary and it is being made.

Integrating the electricity power produced by renewable sources, which are widely distributed in geographical areas, into the existed power system is a great challenge [4]. Expansion of power transmission system is inevitable.

Establishing high voltage direct current (HVDC) overhead lines is a hot topic, due to its desirable power transmission capacity over long distance [5, 6]. However, the application of HVDC needs years of research to overcome the technical issues before put into use in large scale [7]. As a result, building new overhead lines with alternative voltage is still a major option for the power transmission system. Taking the situation in Europe as an example, more than 28000 km of 400 kV overhead lines are in demand [8].

Building a new overhead line, which is a complicated structure usually spanning hundreds of kilometers and shall be well blended into the environment, is never a simple and easy-to-achieved task [9]. Within the past century, we have summarized substantial technical experience for installing overhead lines, which are helpful guidelines in the design and operation of overhead lines [9]. Apart from the technical parts in installing new overhead lines, their impacts on the environment are also a main concern to Transmission System Operators (TSO).

The first symposium regarding "Overhead lines impacts on environment and vice versa" was organized by CIGRE in 1981. CIGRE TB 147 reviewed various impacts of overhead lines on the environment, including the impact of lines corridors on land use, visual impact, construction and maintenance impact [10]. In order to get permission from the government and acceptance from the residents living nearby, TSOs have given considerable thought to mitigating these issues. They believe a new design of transmission pylons is the basis to improve impacts of overhead lines on the environment [9].

1.2.2 Aesthetical Overhead Transmission Pylons

The first transmission pylon, in a simple shape of single wooden pole, was established in the end of 19 century for a 15 kV AC transmission line [9]. Over decades' development, the transmission voltage level reaches up to 1000–1200 kV, which

Fig. 1.1 Conventional lattice pylons **a** a 400 kV double circuit tower in UK, **b** a 500 kV single circuit tower in USA

requires taller and stronger transmission pylons [11, 12]. Nowadays, transmission pylons are made from wood, concrete and steel. For higher voltage levels (>220 kV), steel lattice pylons are dominant. The steel lattices pylons, with a huge and complicated structure, are ubiquitous and have not experienced big changes for several decades, as shown in Fig. 1.1.

As increasing concerns are expressed in the environmental impact of overhead lines, the public is showing growing opposition to the construction of new lattice pylons, whose structure are regarded being ugly for the surrounding landscape.

As a consequence, more aesthetical designed pylons have emerged in different countries, which are featured by compactness and simplicity [13]. For instance, Eltra, an old TSO in Denmark employed a new design of transmission pylon for a new 400 kV overhead line between the cities of Aalborg and Aarhus, in order to avoid visual noise and interference to the landscape [14]. The new pylon, shown in Fig. 1.2, has few elements compared with the conventional Donau pylon, which is usual in Denmark, thus a simplicity can be achieved [14]. Meanwhile, some architecture design companies have proposed new transmission pylon configurations, as shown in Fig. 1.3. Although these designs are only concepts for the time being, they indicate a trending for redevelopment of overhead lines, which places more emphasis on the visual impact.

Fig. 1.2 **a** A prototype of new pylon for 400 kV overhead line between Aalborg and Aarhus (in Denmark), **b** A conventional Donau pylon for 400 kV overhead line [14]

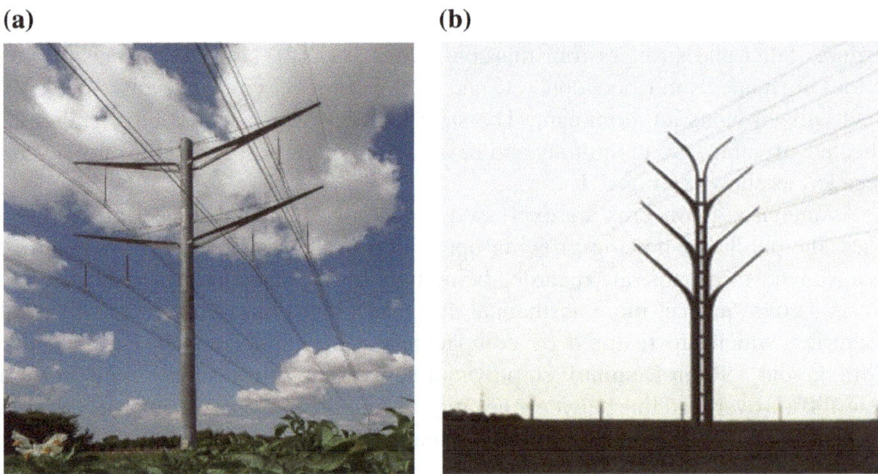

Fig. 1.3 Examples of aesthetical transmission pylon concepts, **a** Eagle pylon configuration proposed by Bystrup company, **b** Flower pylon configuration proposed by Gustafson Porter with Atelier One and Pfisterer

1.2.3 Composite-Based Transmission Pylons

The development of material science, especially the application of Fiber Reinforced Plastic (FRP) composite materials contributes to the expansion of those aesthetical designed pylons. Fiber Reinforced Plastic (FRP) materials have the features of good electrical insulation property as well as the high mechanical strength. Thus, FRP composite-based transmission pylons can fulfill the requirements of compactness and electricity transmission capacity.

A pylon based on Fiber Reinforced Plastic (FRP) usually comprises commercial components based on composite materials, such as composite insulators, composite cross-arms and a single pole either made of steel, concrete or FRP. In the North America, composite-based pylons have been investigated back to 1960s [15]. Over the last decade, more researches on the composite-based pylons have been implemented in Europe and other countries. For instance, the Dutch TSO TenneT has launched Wintrack and Wintrack II project and proposed an innovative pylon concept [16], shown in Fig. 1.4. The three designs of pylon in Fig. 1.4a are based on an I-pole body pylon with insulating cross-arms. The cross-arms was planned to be made out of an insulating composite material making the use of separated cross-arms and insulators redundant and achieving a minimization of the overhead line system's visual impact. Three different ideas were introduced for how electromagnetic field can be optimized. On the other hand, the Wintrack II pylon, shown in Fig. 1.4b, has a single concrete pole and composite cross-arms. By using the composite cross-arm without insulator strings, the air clearances on the pylon are able to be optimized and the dimension of the pylon can be reduced. As a result, it takes a small transmission corridor, which is critical considering the fact that available land source is decreasing for power delivery.

A Chinese project also proposes a composite-based pylon either with single pole or double pole for 110 kV lines, which is intended to install in ShenZhen, a city in the coast of Southern China [17]. The new pylon, shown in Fig. 1.5, has single/double

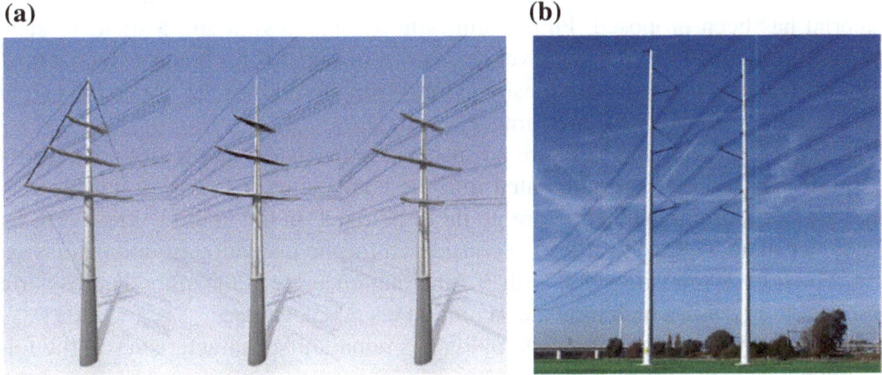

Fig. 1.4 **a** Wintrack Project [13], and **b** Wintrack II pylon [16] by Dutch TSO TenneT

Fig. 1.5 A composite pylon for 110 kV overhead lines in China, **a** with a single pole, **b** with double poles [17]

pole(s) made of glass fiber reinforced polyurethane. The cross-arm of the pylon is composed of two commercial composite insulators.

More projects of composite-based transmission pylons are introduced in [13]. With the concerns of overhead lines' visual impact and FRP materials' favorable characteristics in mind, the composite-based pylon is a promising solution for the next generation of transmission pylons.

1.3 Introduction of Power Pylons of the Future Project

In 2011, a competition was held in UK for the new design of the next generation of power pylons in which T-pylon design by Bystrup architects was chosen. In the next development of T-pylon concept, a novel pylon design idea based on fully composite material has been proposed. Fully composite pylon, T-Pylon and a standard steel lattice tower at 400 kV voltage level are shown in Fig. 1.6 for better visualization. Figure 1.6 shows that fully composite pylon is roughly half of the lattice tower and has a more appealing visual appearance.

It can also be seen that there are no suspension insulators on the fully composite pylon design and they are integrated into the composite cross-arm design namely unibody cross-arm. Conductors are going to be fixed on the unibody cross-arm by conductor clamps. Shed profiles are considered for the insulation of unibody cross-arm. The fully composite pylon with 30° inclination angle of the unibody cross-arm is shown in Fig. 1.7 which indicates a unique visual appearance.

In summary, the fully composite pylon has remarkable characteristics in the following perspectives [1, 8]:

- The pylon carries two circuits of 400 kV AC lines.

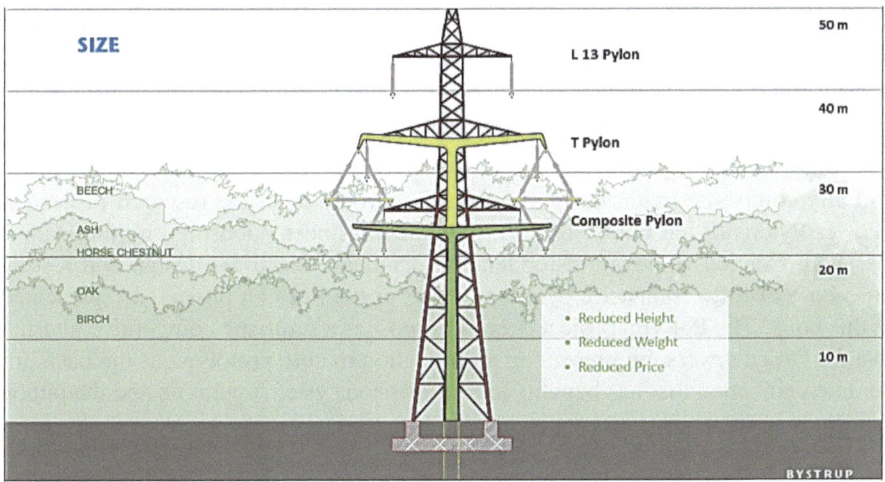

Fig. 1.6 Height of a standard lattice tower, T-Pylon and Fully Composite Pylon and their appearance in landscape [8]

Fig. 1.7 Visual appearance of fully composite pylon with 30° inclination angle in landscape [1]

- The pylon has an integrated cross-arm without insulator strings. In another word, the one piece cross-arm itself is an insulator, whose function is isolating three phase conductors electrically and providing mechanical support at the same time.
- To address the large moment due to the one-piece cross-arm, the cross-arm insulator is made from composite materials that have the feature of light weight. Thus, the pylon is a composite-based transmission pylon. The pylon pole can be made either from composite material, concrete or steel.
- All the conductors are attached to the unibody cross-arm directly by specific conductor clamps. Distance between phases on the pylon equals to the arcing distances.
- Two shield wires are fixed by clamps at the tips of the unibody cross-arm. As a result, the shielding angle for the pylon is negative.
- With the special and unique configuration, the fully composite pylon has a height of 22.5 m only, which is much lower than that of conventional lattice pylons for the same voltage level, as Fig. 1.6 shows.

- The pylon requires a smaller transmission corridor compared with its counterparts.
- A 250 m span length is applied with the pylon.
- With the compact configuration, the pylon can more easily fit into the landscape and impose less visual impact to residents nearby. Thus, it's expected to get acceptance more easily from the public.

Fully composite pylon design concept is introduced in the research project of "Power Pylons of the Future (PoPyFu)" which has been sponsored by the Danish Research Council (Innovationsfonden) in Denmark. Electrical design and testing procedure of fully composite pylon, as a part of the PoPyFu project, are presented in this book. The PoPyFu project overall purpose is a futuristic competitive alternative for lattice towers, by presenting a production mature prototype as the basis for commercialization that has benefits to transmission system operators and the public especially in terms of visual appearance.

1.4 Challenges and Research Objectives

The upcoming challenges in the design of fully composite pylon are addressed in this section. One of the challenges is to select the most suitable Fiber Reinforced Plastic (FRP) material in the application to the composite cross-arm. Pure electrical tests are conducted on candidate Fiber Reinforced Plastic (FRP) material samples. Critical electrical properties, including partial discharge (PD) activities and dielectric properties of the material are tested and evaluated. According to operational conditions, the cross-arm is loaded with multiple stresses simultaneously, especially electrical stress and mechanical forces, which lead to material aging and conse-quently deteriorate the composite cross-arm's lifetime. Thus, electrical and mechanical combined test is conducted on the Fiber Reinforced Plastic (FRP) material samples, and their behaviors under combined stresses is investigated.

Another major challenge in the electrical design of fully composite pylon is the lightning shielding of fully composite pylon which is due to the location of shield wires on the cross-arm. Therefore, the second objective is to investigate the feasibility and effectiveness of lightning protection system for the fully composite pylon. The preliminary concept of fully composite pylon comprised of two different designs for the lightning protection of the pylon including fully composite pylon with and without shield wires (Fig. 1.8). A fully composite pylon without shield wires was the first alternative which would utilize surge arresters to protect the pylon against lightning strikes. The idea was based on mounting surge arresters inside the hollow unibody cross-arm and connecting them to the earthing system through a bare ground wire inside the I-pole body. However, it is discussed in [18] that internally mounting surge arresters within the unibody cross-arm are not feasible due to insufficient free space inside the cross-arm, dimensions of surge arresters, electric field limitations, safety clearances, maintenances and economical and mechanical aspects. Externally mounting of surge arresters on the fully composite pylon also disfigures the visual

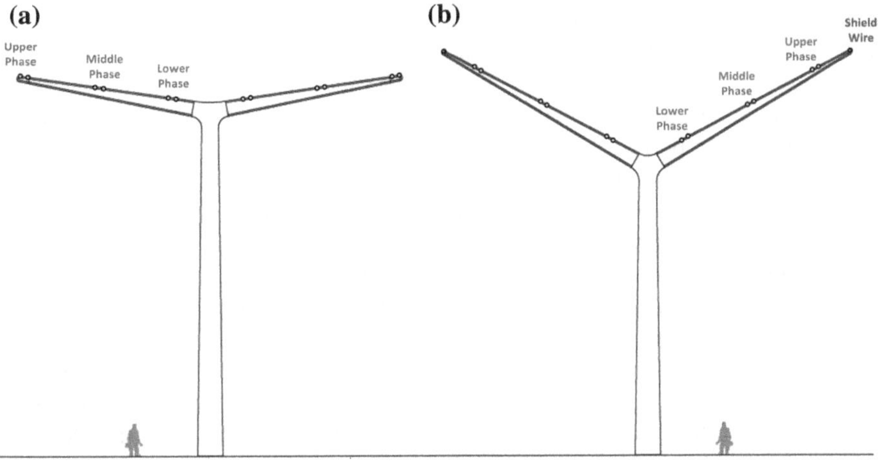

Fig. 1.8 Fully composite pylon concept, **a** without shield wires, **b** with shield wires

appearance of the pylon. For the reasons, the concept of fully composite pylon without shield wires was discarded.

The second alternative for the lightning protection system is the installation of two shield wires at both tips of the pylon, which is a feasible and reliable solution. The objectives are evaluation of lightning performance of fully composite pylon by conventional methods (e.g. EGM) and determination of effectiveness of assigned shielding angle for the pylon. Electro-geometric model (EGM) is widely used for predicting the lightning shielding performance in the design of overhead lines in industry. However, some researchers have pointed out in some cases the real shielding failure performance of overhead lines in service differs from that predicted by electro-geometric model (EGM), especially in the case of ultra-high voltage (UHV) lines. This is because electro-geometric model (EGM) only relates the striking distance to lightning current and ignores physical process of a lightning strike. Thus, the theoretical analysis by electro-geometric model (EGM) needs to be verified by practical experience, or more efficiently, by experimental methods. In order to verify the results by electro-geometric model (EGM) methods and evaluate the lightning shielding performance of the fully composite pylon which has an unusual negative shielding angle, scale model tests have been performed.

Furthermore, since the fully composite pylon is made of composite materials, the question is how ground potential access can be reached for the shield wires. In this regard, two options are proposed as follows:

- Utilization of a grounded cable through the I-pole body and unibody cross-arm to connect shied wires to the earthing system
- Utilization of an external ground connector (e.g. bare conductor) from outside the pylon between the earthing system and shield wires

The main question is the influence of internal ground cable/external ground connector on the electric field and potential distribution around and inside the unibody cross-arm of fully composite pylon. This major challenge is followed in [19, 20] by performing numerous finite element modeling of fully composite pylon and therefore, electric field performance of the pylon is investigated and evaluated and the results are discussed in the next chapters. The electric field performance of conductor clamps (for the connections between conductors and unibody cross-arm) is also investigated based on these finite element analyses.

As operational experience of composite insulators indicates, water droplets induced corona activities on the polymer weather sheds of the composite insulator damage its hydrophobicity, leading to its aging in the long-term running. Investigation on water droplets-induced corona activities is of importance in the verification of composite cross-arm's design. With this in mind, the water droplets induced corona activities on a fully composite cross-arm segment is investigated and the electrical field distribution around the composite cross-arm surface is evaluated. Additionally, the inclined angle ($\theta_{cross-arm}$) of the cross-arm may have effects on the water droplets-induced corona discharge from the cross-arm surface. Thus, water induced corona activity on the cross-arm surface with various inclined angles is investigated.

Another important research objective is the dimensioning of fully composite pylon which needs to be specified for insulation design on the unibody cross-arm. In order to determine appropriate electrical clearances on the pylon, an insulation coordination study is needed to be performed. This study is required to evaluate the electrical behavior of the pylon against power frequency voltages and switching/lightning impulse overvoltages. The outcome of the study is determination of required air clearances on the pylon between two phases' as well as phase and shield wire [21]. Estimating required creepage distances for the insulation of unibody cross-arm and allocating proper shed profiles on the sheath are other objectives that are investigated.

In a traditional insulator design, phase-to-ground air clearance (between live-end and grounded-end) is used for the calculation of phase-to-ground creepage distance of the insulator. However, on the unibody cross-arm of fully composite pylon, in addition to phase-to-ground creepage distance, there is also another creepage distance which is phase-to-phase creepage distance between two adjacent phases. Therefore, a different approach is needed to calculate the required phase-to-phase creepage distance on the unibody cross-arm.

Environmental effects of fully composite pylon including audible noises, radio interferences, corona losses and electromagnetic emissions at the right-of-way (ROW) of the line are also needed to be calculated and evaluated.

The main researches in this book can be summarized as follows:

- Dimensioning of fully composite pylon by determining required air clearances on the pylon
- Determining the most suitable fiber-reinforced plastic material as core of the composite cross-arm and the most suitable material manufacturing method
- Designing insulation for the unibody cross-arm by calculating required creepage distances on the cross-arm and allocating appropriate shed profiles on shed housing

- Investigating the influence of internal ground cable/external ground connector on the electric field and potential distribution around and inside the unibody cross-arm of fully composite pylon
- Evaluating electric field performance of fully composite pylon with different conductor clamp designs
- Validating the electric field distribution on the composite cross-arm surface in wet conditions
- Investigating the feasibility and effectiveness of lightning protection system for the fully composite pylon
- Validating the cross-arm inclined angle from the ground plane
- Evaluating environmental effects of fully composite pylon including audible noises, radio interferences, corona losses and electromagnetic emissions.

1.5 Outlines of Book

This book is divided into nine chapters and covers the electrical design and testing process of fully composite pylon.

Chapter 1: In this chapter, the explanation of new design trending of pylons for future overhead lines is presented. It is followed by an introduction of the Power Pylons of the Future (PoPyFu) project, which aims to propose and verify the electrical design of an innovative fully composite transmission pylon. In this regard, design challenges and objectives are given in this chapter.

Chapter 2: This chapter introduces the basic knowledge of fiber reinforced plastic (FRP) materials, including their composition and important properties. Additionally, the state of the art review of the fiber reinforced plastic's application in power transmission pylons is introduced. Based on the technical experience and de-sign guidelines of composite components in transmission pylons, electrical experiments on candidate fiber reinforced plastic samples manufactured by different methods are introduced. Results including dielectric properties and partial discharge (PD) activities are used as indicators in the selection of fiber reinforced plastic materials for the composite cross-arm core. Meanwhile, effects of mechanical loading on materials' electrical behaviors, especially partial discharge behaviors, are also investigated.

Chapter 3: This chapter is related to the insulation coordination study to specify required electrical clearances on the fully composite pylon. The insulation coordination procedure to determine minimum required air clearances on the fully composite pylon is described in the chapter. The obtained results for the basic design of fully composite pylon are presented and the derived internal and external clearances are given.

Chapter 4: Electric field performance of fully composite pylon is presented in this chapter. The design of insulation for the unibody cross-arm is explained in more details and the obtained results are presented. Based on the electric field and potential distribution results, the basic design of fully composite pylon is updated and the required modifications (especially for the conductor clamp design) are applied and further finite element analyses are carried out to achieve a successful design of fully composite pylon.

Chapter 5: This chapter investigates water-induced corona activities on the unibody cross-arm phase-to-phase section. By analyzing the corona inception voltage/electric field and partial discharge (PD) intensities, the electric field distribution on and around the cross-arm is evaluated. Consequently, suggestions for design improvement of the composite cross-arm based on test results have been given.

Chapter 6: Lightning shielding performance of the pylon against vertical lightning strikes is evaluated. Conventional electro-geometric (EGM) method is reviewed and a modified EGM method is introduced for the application of fully composite pylon. The preliminary assigned shielding angle of $-60°$ for the fully composite pylon is assessed and its shielding failure rate (SFR) and shielding failure flashover rate (SFFOR) are calculated. The lightning shielding behavior of fully composite pylon against possible horizontal lightning strikes is also investigated using rolling sphere method (RSM).

Chapter 7: In this chapter, experimentally evaluation of shielding failure behavior of an overhead line composed of fully composite pylons is presented. In order to verify the electro-geometric model (EGM) results, a scale model test is applied to evaluate the lightning shielding probability in the pylon. Moreover, effects of the cross-arm inclined angle on the lightning shielding performance are investigated.

Chapter 8: Environmental effects of fully composite pylon are investigated in this chapter. The surface gradients on phase conductors are used for the computation of audible noises based on empirical methods. In the same way, the radio noises are calculated by semi-analytical and empirical approaches and evaluated. Corona losses are calculated and finally, electromagnetic emissions are calculated and the required right-of-way (ROW) width at both sides of pylon is determined.

Chapter 9: The conclusion and summary of book together with future challenges are presented in the last chapter.

References

1. H. Skouboe, M.H. Mikkelsen, C.L. Bak, et al., The composite pylon, in *CIGRE Session*, August 2018, Paris, France (2018), p. B2-308
2. T.K. Sørensen, J. Holboell, Composite materials in overhead transmission lines present and future solutions, in *Nordic Insulation Symposium (NORD-IS)* (2009)
3. Petroleum British, BP statistical review of world energy, June 2015, http://www.bp.com
4. L. Brid, M. Milligan, D. Lew, *Integrating variable renewable energy: challenges and solutions*. National Renewable Energy Laboratory (2013)
5. E.W. Kimbark, *Direct Current Transmission*, vol. 1 (Wiley, 1971)
6. T.J. Hammons, D. Woodford, J. Loughtan et al., Role of HVDC transmission in future energy development. IEEE Power Eng. Rev. **20**(2), 10–25 (2000)
7. M.H. Okba, M.H. Saied, M.Z. Mostafa, et al., High voltage direct current transmission—a review, part I, in *Energytech* (IEEE, 2012), pp. 1–7
8. BYSTRUP Power Pylon Design, www.powerpylons.com
9. CIGRE, *CIGRE Green Book on Overhead Lines* (Paris, 2014)

10. CIGRE WG14, Environmental concerns, procedures, impacts and mitigations, in *CI-GRE Report 147* (2000)
11. Y. Hu, K. Liu, T. Wu, et al., Key technology research and application of live working technology on EHV/UHV transmission lines in China, in *International Conference on Power System Technology (POWERCON)*, Chengdu, China, Oct 2014, pp. 2299–2309
12. A.R. Reddy, P.C. Sekhar, Estimation of switching surge flashover rate for a 1200 kV UHVAC transmission line, in *7th International Conference on Power Systems (ICPS)*, Pune, India, Dec 2017, pp. 358–363
13. T.K. Sørensen, Composite based EHV AC overhead transmission lines. Ph.D. thesis, Technical University of Denmark, 2010
14. H. Oebro, E. Bystrup, K. Krogh, et al., New type of tower for overhead lines, in *CIGRE Session B2* (2004)
15. M. Sarmento, B. Lacoursiere, A state of the art overview: composite utility poles for distribution and transmission applications, in *Transmission and Distribution Conference and Exposition: Latin America*. Caracas, Venezuela, Aug. 2006, pp. 1–4
16. TenneT. Wintrack II high-voltage pylons, www.tennet.eu
17. H.M. Li, S.C. Deng, Q.H. Wei, et al., Research on composite material towers used in 110 kV overhead transmission lines, in *International Conference on High Voltage Engineering and Application*, New Orleans, USA, Oct 2010, pp. 572–575
18. T. Jahangiri, C.L. Bak, F.F. Silva, B. Endahl, A state of the art overview: EHV composite cross-arms and composite-based pylons, in *The 19th International Symposium on High Voltage Engineering*, Pilsen, Czech Republic (2015)
19. T. Jahangiri, C.L. Bak, F. M.F. da Silva, B. Endahl, Electric field and potential distribution in a 420 kV novel unibody composite cross-arm, in *Proceedings of the 24th Nordic Insulation Symposium on Materials, Components and Diagnostics* (2015), pp. 111–116
20. T. Jahangiri, Q. Wang, C.L. Bak, F. M.F. da Silva, H. Skouboe, Electric stress computations for designing a novel unibody composite cross-arm using finite element method. IEEE Trans. Dielectr. Electr. Insul. **24**(6), 3567–3577 (2017)
21. T. Jahangiri, C.L. Bak, F.F. da Silva, B. Endahl, Determination of minimum air clearances for a 420 kV novel unibody composite cross-arm, in *2015 50th International Universities Power Engineering Conference (UPEC)* (2015), pp. 1–6

Chapter 2
Fiber Reinforced Plastic (FRP) Composite Selection for the Composite Cross-Arm Core

2.1 Fiber Reinforced Plastic (FRP) Composites

Fiber Reinforced Plastic (FRP) composite material becomes more and more popular in many industrial applications, including marine, aerospace, civil engineering, electrical power engineering (especially in wind turbines) and automotive areas [1–5].

Fiber Reinforced Plastic (FRP) is mainly comprised of a kind of fiber, as the 'skeleton' of the material matter, and a kind of polymer as the matrix, which usually takes 30–70% of the volume of the composite material [6]. In principle, fibers have high tensile strength. However, they are not able to withstand high sheer, bending and compressive forces. Also, they are usually vulnerable to chemical corrosions. In order to overcome the shortcomings of single fibers, polymer matrix is utilized to bond loose fibers together, forming a figurate composite material. In a word, in Fiber Reinforced Plastic (FRP) materials, fibers provide strength whereas polymers provide shape and protection [7]. Apart from these two main parts in Fiber Reinforced Plastic (FRP) composites, a slight amount of additives, such as anti-ultraviolet components, fillers that can change the material's color and fire inhibitors, etc. are added to enhance the material properties according to various requirements in different applications [8, 9]. The properties of a Fiber Reinforced Plastic (FRP) composite mainly depend on selection of fibers, polymers and how they interact with each other with different manufacturing methods [10].

2.1.1 Fibers

The most common used fibers in Fiber Reinforced Plastic (FRP) composites are glass fiber, carbon fiber, aramid fiber and basalt fiber [11, 12].

Glass fibers are the most widely used fibers among Fiber Reinforced Plastic (FRP) materials, since they possesses the advantages of good mechanical strength, desirable

T. Jahangiri et al., *Electrical Design of a 400 kV Composite Tower*, Lecture Notes in Electrical Engineering 557,
https://doi.org/10.1007/978-3-030-17843-7_2

anti-corrosion capacity and good heat-resistance property [2, 12]. In addition, glass fiber is a dielectric, which is electrical insulating. However, glass fiber is subject to creep with long-lasting loading and to degradation in alkaline conditions. Glass fibers have different types, mainly including E-glass, E-CR-glass, S-glass, R-glass and T-glass [2].

Carbon fiber has a high carbon content of more than 90%. It has the characteristics of low density, high mechanical strength and good anti-corossion property [12]. Compared with glass fibers, carbon fibers have higher elastic modulus, thus a higher stiffness. Consequently, it is usual in applications which requires high ratio of stiffness to weight, such as in aerospace field and military field. However, the carbon fiber is electrical conductive, restraining its availability in electrical equipment.

Aramid fiber, apart from its characteristics of outstanding strenght to weight property, good heat resistance and high elastic modulus, it also has advantages of impact resistance and anti-acid/alkaline property [12]. It is electrical insulating thus it is possible to apply to electrical systems. However, aramid fiber is easy to absorb moisture. Also, its production cost is higher compared with glass fiber.

Basalt fiber, as its name indicates, is made from basalt. As with glass fiber, it possesses advantages of good dielectric strength, anti-corrosion property and good heat resistance [12]. What's more, it is an environmental-friendly material, since it produces few waste during manufacturing process, which can easily degrade. Consequently, basalt fiber is regarded as a promising alternative for future fiber applications.

In electrical power system applications, E-glass fiber is most common when using Fiber Reinforced Plastic (FRP) composites, due to its large production rate, relatively low cost and good electrical as well as mechanical strength [13–15].

Apart from fiber types, the forms of fibers also have great effect on the performance of Fiber Reinforced Plastic (FRP) composites [2]. In principle, the fiber forms are divided into short fibers, long fibers, woven fibers and chopped fibers. Fiber form varies corresponding to different applications.

2.1.2 Polymers

Polymers are divided into two categories: thermoset polymers and thermoplastic polymers [16]. The former indicates a polymer cured from liquid or soft solid, which will not melt again when exposed to high temperature [16]. On the contrary, thermoplastic polymer indicates a cured polymer that will melt when exposed to high temperature [16]. Since a transmission pylon may be installed in high temperature areas, such as in a desert, a thermoset polymer is preferable when considering the application of Fiber Reinforced Plastic (FRP) in transmission pylons. Thus, the polymers discussed in the present thesis are all thermosets.

In the final composite material bulk, polymer works as binding agent that bonds loose fibers together. Meanwhile, it transfers stresses to fibers. In general, polymers have the advantages of lower density compared with metals, good electrical insulating property and anti-corrosion capacity. Meanwhile, they have a weak mechanical strenght which can be enhanced by fibers in Fiber Reinforced Plastic (FRP) composite materials. The most commonly used thermoset polymers in Fiber Reinforced Plastic (FRP) are polyester, vinyl ester, polyurethane and epoxy [12].

Polyester is a preferable insulation material in the application to capacitors, transformers, cables etc., since it has high relative permittivity, high electrical strength and good thermal stability [17–19]. Among the polymers mentioned above, polyester has the lowest cost and moderate mechanical properties. However, it is sensitive to Ultra Violet (UV) degradation.

Vinyl ester has higher mechanical strength than polyester. Due to its high chemical corrosion resistant and moisture resistant properties, it is widely used in the marine industry [20]. However, it is sensitive to heat and its production cost is high.

Polyurethane has high flexibility, which means its mechanical properties can be adjusted within a wide range through adjusting ratios in raw materials [21]. At the same time, it has admirable chemical resistant properties. However, polyurethane's cost is almost 1.5 times more expensive than vinyl ester.

Epoxy has high mechanical and thermal properties. Meanwhile, it is moisture- and heat- resistant. Epoxy reinforced by glass/carbon fibers are stronger and more heat resistant than other Fiber Reinforced Plastic (FRP) composites. And it is electrical insulating thus employed as insulation material in electric power system equipment, such as transformers, switchgears and insulators [10, 22].

2.1.3 Manufacturing Methods

There are different ways to manufacture Fiber Reinforced Plastic (FRP) composites, based on types, sizes, configurations and quality demands of the final Fiber Reinforced Plastic (FRP) products [23]. Common manufacturing methods for Fiber Reinforced Plastic (FRP) composite materials are summarized in [7, 10] and briefly introduced below.

Hand lay-up moulding, also called contacting molding, is a manufacturing method for Fiber Reinforced Plastic (FRP) composites in the earliest period. The process is specified by impregnating a fiber fabric into a layer of polymer resin (matrixs), which is deposited in a mold beforehand. Then the fabric is pressed by a roller or rabbler, in order to force the polymer resin fully infiltrate the fabric and squeeze out the air as much as possible. After that, a second resin and fabric are added and this process is repeated until the desired thickness is acquired. Hand lay-up moulding requires simple equipment and costs low. And it is not limited by size and shape of the product. However, it is also featured by low efficiency, long production cycle and relative low quality. For applications where high reliable performance is required, this method is inadvisable.

In order to overcome the shortcomings of the hand lay-up moulding, a more reliable compression moulding method is applied. In stead of stressing fiber fabrics into polymer resins by manual operation, the compression moulding method utilizes high pressure and high temperature, which mixture these two components more thoroughly.

Unlike compression moulding, vacuum moulding method utilizes a vacuum condition to apply pressure on fabric and resin mixtures. After a layer of mixture is deposited in the mold, a plastic film is laid on it and a vacuum pump extracts air under the film. As a result, the atmospheric pressure is applied to the mixture and air or disable gases in the resins are squeezed out.

Vacuum assisted resin transfer moulding method is developed from vacuum moulding. With this method, fiber fabrics are placed in a mold sealed by a plastic vacuum bag. A vacuum pump connecting to the mold drives the liquid polymer resin infusing into the fabrics. During this process, air or dissable gases between laminates of fibers are squeezed out. Vacuum assisted resin transfer moulding method is widely used in the marine industry.

Pultrusion indicates fiber bundles are pulled through a resin impregnator and a forming die in sequence. This method is widely used for manufacturing tubular products, bars/rods, square products and so on.

Filament winding is a manufacturing method that winds fiber bundles, which have been impregnated in polymer resins throughly in advance, onto a rotating mandrel. Winding patterns, fiber angle and numbers can be controlled by computer automatically.

In applications to power transmission grids, the last two methods mentioned above are most commonly used to manufacture electrical components [7, 10, 19].

2.2 Application of Fiber Reinforced Plastic (FRP) Composites to Transmission Towers

The application of FRP composites to transmission towers mainly takes advantages of its good dielectric strength, high ratio of weight to mechanical strength and anti-corrosion property. Commercialized components making from Fiber Reinforced Plastic (FRP) composites in transmission towers include insulators, cross-arms and tower bodies.

2.2.1 Composite Insulators

Composite insulators for overhead lines were first introduced in 1960s in Europe and USA [24, 25]. Later on, they have been obtaining increasingly acceptance for a large range of voltage levels all over the world [26–28]. Compared with their ceramic and glass counterparts, composite insulators have advantages of light weight, high mechanical strength, good anti-contamination performance, easy installation and competitive cost [27, 29, 30].

As Fig. 2.1 shows, a composite insulator is usually comprised of a Fiber Reinforced Plastic (FRP) solid rod/hollow tube, polymeric housing and metallic end

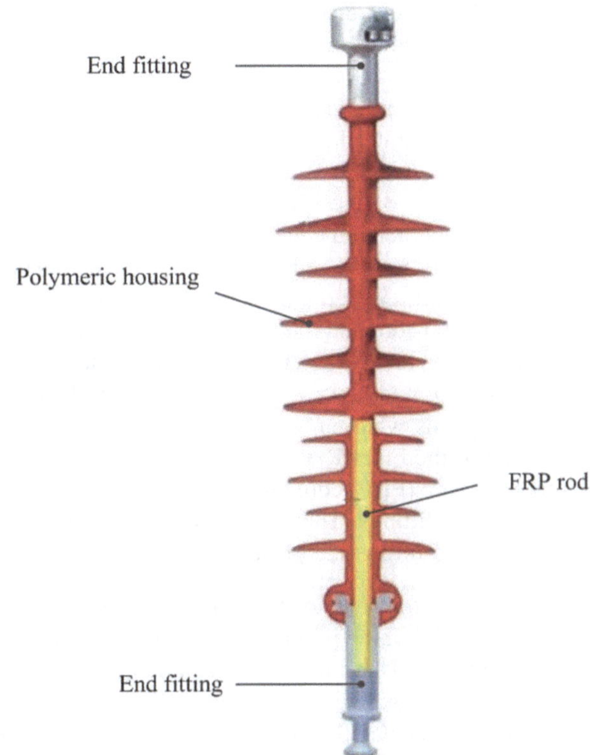

Fig. 2.1 A typical composite line insulator

fittings [31]. There are two main functions of the Fiber Reinforced Plastic (FRP) rod/tube. One is to deal with mechanical load from conductors with consideration of wind and icing, while the other one is to provide insulation between phase conductors and the tower body. The polymeric weather sheds and sheath are applied to provide enough creepage distance and protect the Fiber Reinforced Plastic (FRP) core from aging and degradation. The metallic end fittings are critical for sealing the weather sheath and also necessary to connect phase conductors.

Despite composite insulators provide many benefits compared to their counterparts—ceramic and glass insulators, they may fail in service due to exposure to mechanical, electrical, thermal and environmental effects simultaneously [7].

Failure modes of composite insulators have been surveyed by different institutes based on service experience of more than one million composite insulators [26, 28, 32]. Although differences exist in proportions of composite insulator failure modes in different regions, the main modes can be summarized as following [7, 26, 28, 32]:

- Brittle fracture, which was reported as the main failure cause of composite insulators in USA. Brittle fracture indicates a sudden mechanical failure of a composite insulator core. The occurrence of brittle fracture is caused by electrical discharge in the Fiber Reinforced Plastic (FRP) core or at the interface of Fiber Reinforced

Plastic (FRP) core and the polymeric housing, with the presence of water. Due to its catastrophic results, many researchers have performed investigation on this topic [26, 33].

- Flashover, which indicates an electrical failure of the composite insulator that can no longer separate different voltage potentials as it should do. Flahover along a composite insulator is caused by various reasons, such as lightning strokes, switching overvoltages, severe contaminations, etc. [7, 28].
- Flashunder, which indicates a flashover under the polymeric housing, either along or in the Fiber Reinforced Plastic (FRP) core. A flashunder may be caused by bad seal of end fittings, loose adhesion of polymeric housing to the Fiber Reinforced Plastic (FRP) core and the formation of punctures on the polymeric housing.
- Insulator puncture, which is classified as a fatal accident since it often causes complete explosion of housing thus falling of conductors [28]. The insulator puncture often happens when an overvoltage is present. A better adhesion of polymeric housing to the Fiber Reinforced Plastic (FRP) core and an enhanced Fiber Reinforced Plastic (FRP) core quality can reduce the possibility of insulator puncture.
- Loss of end fittings. The Fiber Reinforced Plastic (FRP) core may be pulled out from the end fitting due to bad sealing with the presence of mechanical load.

Mechanisms that lead to above-mentioned composite insulators failure have been discussed in [7], which will not be introduced in details here.

In order to avoid the composite insulator failure and make it more reliable, different institutes have issued widely accepted standards, specifying design rules, validation methods and acceptance criteria for composite insulators.

2.2.2 Composite Cross-Arms

Traditionally, a cross-arm is utilized to separate conductors from the tower body, insuring an enough air clearance between phase conductors, which are at high voltage potential, and the grounded tower body. In general, a traditional cross-arm in power transmission towers for voltage higher than 220 kV is made from steel and thus conductive. A traditional cross-arm can be regarded as the extension of a tower body.

Back to 1960s, composite-based cross-arms were firstly introduced for a 66 kV transmission tower in Japan [34], shown in Fig. 2.2. The proposal of the composite cross-arm aimed to eliminate flashover from one phase conductor to the adjacent one, due to wind-swing of the suspension insulator [34]. This primitive Fiber Reinforced Plastic (FRP) cross-arm was simply comprised of two bare Fiber Reinforced Plastic (FRP) bars, forming a ‘V’ shape. One bar was the compression member, while the other one was the tension member in the structure. A suspension type ceramic insulator was connected to the cross-arm. The application of the composite cross-arm reduced the required air clearances in the tower. However, the bare Fiber Reinforced

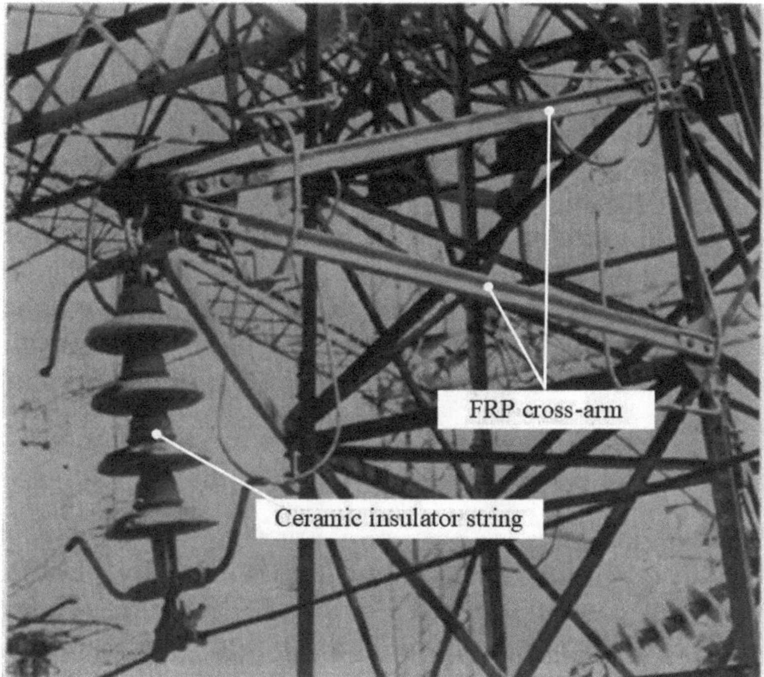

Fig. 2.2 A Fiber Reinforced Plastic (FRP) cross-arm proposed for Kyronan line operating at 66 kV in Japan [34]

Plastic (FRP) bars were vulnerable to electrical discharges on its surface, especially when contamination was presented.

Nowadays, a composite-based insulating cross-arm has been accepted as a solution for compact transmission lines, which take less transmission corridor and reduce the impact of overhead lines to the enviroment [31, 35, 36]. Additionally, the application of a composite/composite-based cross-arm makes the non-use of insulator strings possible, or at least the length of the insulator string is able to be shortened to a large degree. As a results, it reduces the probability of flashover accidents caused by insulator swing due to strong wind.

Now, a typical composite cross-arm is comprised of several composite insulators, as Fig. 2.3 shows, where the V-shaped composite cross-arm is comprised of a compression composite insulator and a tension composite insulator. Due to higher requirements on the mechanical strength of a composite cross-arm, it is larger in size compared with commercialized composite insulators [37]. More insulators are involved in higher voltage levels to address bigger mechanical forces and higher requirement for reliability.

As introduced in [39, 40], a composite cross-arm is proposed in order to compact and upgrade transmission lines. As it is shown in Fig. 2.4, the cross-arm is comprised of four commercialized composite insulators, two of which takes tension forces while

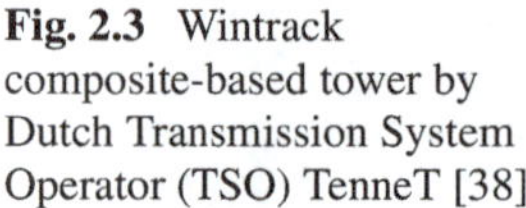
Fig. 2.3 Wintrack composite-based tower by Dutch Transmission System Operator (TSO) TenneT [38]

the rest takes compression forces from conductors. In this case, phase conductors are directly connected to the end of the composite cross-arm through special noses, eliminating the application of long insulator strings.

While in [41], a similar structure of composite cross-arm, which has six insulators, is developed, shown in Fig. 2.5. Since the cross-arm is designed for Extra High Voltage (EHV) lines, an extra insulator string is applied to the end of the cross-arm for connecting phase conductors. However, it is much shorter compared with traditional insulator strings in 750 kV AC lines.

More design guidelines regarding composite cross-arms are introduced in [42].

2.2.3 Composite Tower Poles

Tower body is a strong structure supporting transmission and distribution lines. For the former case, usually steel and sometimes concrete are used, while wood is also a choice for the latter case [31]. Tower bodies made from these materials suffer from different problems: steel tower body is conductive thus the air clearances needed in the tower are relatively large, and the increasing severity of contamination leads to a even bigger tower dimension [19]. Concrete body is heavy thus the tower footing is quite large and the installation cost is substantial [19]. Wooden tower body is vulnerable to water corrosion and damages from birds and worms [19].

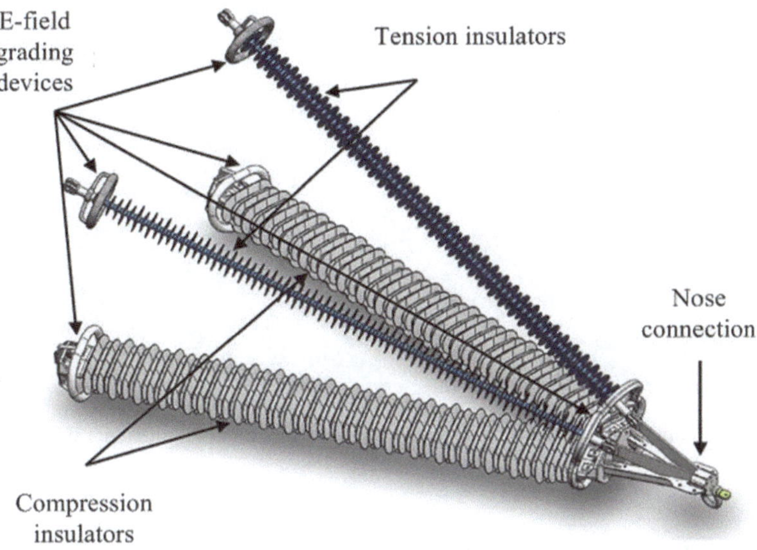

Fig. 2.4 A composite cross-arm proposed for existing and new overhead lines operating at 132 kV and above [40]

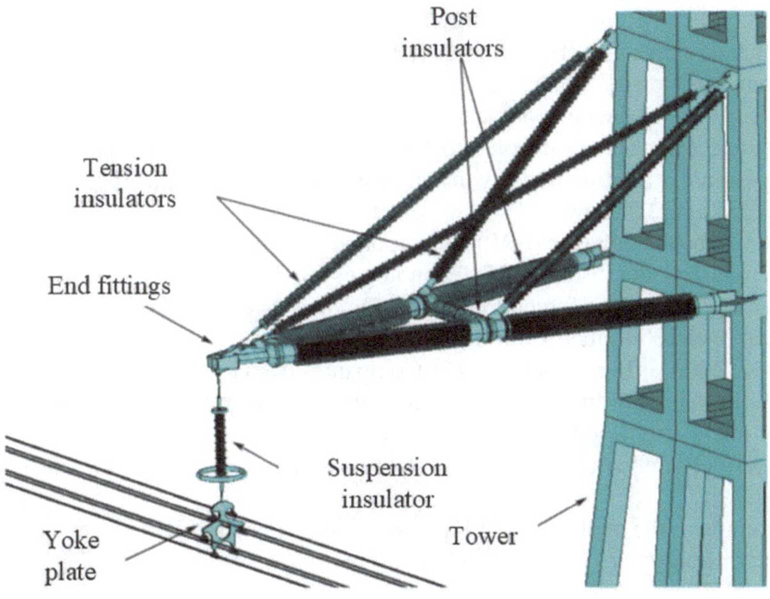

Fig. 2.5 A composite cross-arm proposed for 750 kV AC lines in China [41]

Fig. 2.6 A composite tower for 110 kV overhead lines in China with a single pole (left); with double poles (right) [45]

Compared with its counterparts, composite tower pole benefits a lot from Fiber Reinforced Plastic (FRP) materials, from which it is made. Properties of Fiber Reinforced Plastic (FRP) make the composite tower more compact, much lighter and stronger, corrosion resistant and immune to birds/worms damages [43].

The first Fiber Reinforced Plastic (FRP) tower, which is a monopole, is installed on the island of Maui in Hawaii in 1960s, aiming to overcome the shortcomings of wood and steel towers which are either vulnerable to degradation or to salt moisture corrosion [44]. As the development of material science and manufacturing technology, more researches are devoted to Fiber Reinforced Plastic (FRP) towers in the transmission/distribution lines, which are usually in monopole or H-frame structure as Figs. 2.3 and 2.6 show.

Fiber Reinforced Plastic (FRP) poles applied in distribution/transmission lines are commercialized. For instance, RS Technologies Inc. from Canada is devoting itself to investigating Fiber Reinforced Plastic (FRP) poles for distribution and transmission lines and over 2000 Fiber Reinforced Plastic (FRP) poles have been installed in the field since 2003, as Fig. 2.7 shows [46]. Companies from USA, like Ebert Composites Corporation also produces Fiber Reinforced Plastic (FRP) poles. Those poles are usually made from E-glass fibers reinforced polymers (including polyester, vinyl ester and polyurethane) and are manufactured by pultrusion or filament winding method.

In summary, the application of Fiber Reinforced Plastic (FRP) towers in distribution/transmission lines is attracting increasing interests from all over the world.

Fig. 2.7 Commercialized Fiber Reinforced Plastic (FRP) towers in transmission lines manufactured by RS Technologies Inc., Canada [46]

2.3 Fiber Reinforced Plastic (FRP) Composites in the Fully Composite Pylon

2.3.1 Structure of the Composite Cross-Arm

The configuration of the composite cross-arm in the fully composite pylon proposed in the project Power Pylons of the Future (PoPyFu) is similar to that of a composite insulator, which is composed of three main parts: weather sheds, a Fiber Reinforced Plastic (FRP) core and metallic end-fittings.

The major structure of the cross-arm is a hollow Fiber Reinforced Plastic (FRP) tube, which is surrounded by weather sheds. The tube will be filled either by a kind of gas with a specific pressure or a kind of solid material, which has not been determined yet. As an initial design idea, a grounding cable goes though the Fiber Reinforced Plastic (FRP) tube, which grounds the shield wires. However, this design needs further investigation which will not discussed here.

The Fiber Reinforced Plastic (FRP) tube of the composite cross-arm have two main functions: (1) provide main insulation for conductors at different potentials; (2) withstand the mechanical load from overhead lines considering various cases such as wind, glaciation, etc.

2.3.2 Electrical and Mechanical Effects on the Fiber Reinforced Plastic (FRP) Core

2.3.2.1 Electrical Effect

The Fiber Reinforced Plastic (FRP) core of the composite cross-arm in the fully composite pylon is stressed constantly with nominal power frequency voltage. If the electric field in the Fiber Reinforced Plastic (FRP) core exceeds a critical value, internal partial discharge (PD) may happen from voids, cavities or small cracks in the bulk material. These imperfections are formed during the manufacture process. According to IEC Std. 60270, partial discharge (PD) is defined as

> Localized electrical discharges that only partially bridge the insulation between conductors and which can or cannot occur adjacent to a conductor. Partial discharge (PD) are in general a consequence of local electrical stress concentrations in the insulation or on the surface of the insulation. Generally such discharges appear as pulses having durations of much less than 1 μs [47].

Constant internal partial discharge (PD) is regarded as the main factor leading to aging and degradation of solid insulation in the long-term operation.

First of all, internal partial discharge (PD) produces by-products, such as ozone and solid crystals in the voids, cavities and cracks in the solid insulation. These by-products are harmful to the insulation, since they will react with the insulation material chemically and lead to its degradation in the long-term. If water is presented in the Fiber Reinforced Plastic (FRP) core due to poor sealing at the metallic end fittings, worse case occurs since nitric/nitrous/oxalic acids may be produced in the spot where high electric field magnitude is presented and partial discharge (PD) happens. Fibers in the Fiber Reinforced Plastic (FRP) core material will be eroded by these acids and become brittle. With the mechanical stress from conductors, these brittle fibers will break in the first place. Then the acids will propagate along the broke fibers and lead to corrosion of other fibers. As a result, the other serviving fibers have to sustain more and more mechanical stress until the whole Fiber Reinforced Plastic (FRP) core breaks. This failure model is named as brittle fracture.

Additionally, since the essence of partial discharge (PD) is electrical discharge, there are also ionized particles in the discharge spot which collide with the insulation wall. These collision by ionized particles with high energy damages the insulation material as well. Also, internal partial discharge (PD) in the solid insulation produces heat which leads to thermal degradation.

In a word, the internal partial discharge (PD) at constant power frequency operation voltage is one of the threats to Fiber Reinforced Plastic (FRP) tube in the composite cross-arm. Thus internal partial discharge (PD) should be avoided, which can be fulfilled by two steps: (1) avoid voids, cavities and cracks in the material during manufacture process; (2) control the magnitude of electric field and make sure it is lower than the critical value that can lead to internal partial discharge (PD).

2.3.2.2 Electric Field in the Fiber Reinforced Plastic (FRP) Core

There is not an official standard specifying criteria for electric field magnitudes in Fiber Reinforced Plastic (FRP) core of a composite insulator/cross-arm. However, with years' technical and academic experience, a wide-acceptable criterion is proposed, that the maximum allowable electric field magnitude in the Fiber Reinforced Plastic (FRP) core of the composite cross-arm is 3 kV_{rms}/mm.

In order to make sure the electric field magnitude is lower than the critical value, electric field distribution analysis in the Fiber Reinforced Plastic (FRP) core must be done.

Measuring the electric field in a solid insulation material is impossible, thus a computer-assisted computation with the help of Finite Element Model (FEM) is applied to calculate electric field distribution in the Fiber Reinforced Plastic (FRP) core numerically.

The electric field magnitude computation in the Fiber Reinforced Plastic (FRP) tube of the composite cross-arm in the fully composite pylon will be introduced in details in Chap. 4.

According to calculation in Chap. 4, the maximum electric field magnitude in the Fiber Reinforced Plastic (FRP) core exists below the region where two-bundle phase conductors are located. And the value is around 0.6 kV_{rms}/mm, i.e. 0.85 kV/mm, which is lower than the acceptable criterion.

2.3.2.3 Mechanical Effect

Apart from electrical effect, the composite cross-arm is also loaded by dynamic tensile, compression and shear forces. Behaviors of the composite cross-arm with combined electrical-mechanical stress is of interest.

A Fiber Reinforced Plastic (FRP) composite is an inhomogeneous material, which contains more imperfections than a moulded material, especially at the interface of fibers and the resin. Additionally, a Fiber Reinforced Plastic (FRP) may suffer from fiber detachments and delaminations. When a Fiber Reinforced Plastic (FRP) material is exposed to mechanical stress, shape, size and number of those imperfections in the bulk material may be affected and resultantly electrical behavior of the Fiber Reinforced Plastic (FRP) composite may be affected. In return, electrical discharges, especially internal partial discharge (PD) may happen in a Fiber Reinforced Plastic (FRP) composite, which is not unusual when imperfections exist. Heat, byproducts and high speed ions produced during electrical discharges may degrade the Fiber Reinforced Plastic (FRP) material, which have effects on its mechanical performance.

There are several researches focusing on effects of mechanical stress on electrical behaviors of Fiber Reinforced Plastic (FRP) materials that are utilized as insulation material in electrical apparatus, such as in motor stator, gas-insulated switchgear (GIS), etc. [48–52]. In most of these researches, measuring partial discharge (PD) signals from the Fiber Reinforced Plastic (FRP) material with the mechanical stress applied is the main methodology, as partial discharge (PD) parameters are sensitive to variation of imperfections in the bulk material.

These researches have thrown some light on effects of mechanical stress on electrical performance of Fiber Reinforced Plastic (FRP) composites. In [49], the author has found out that partial discharge (PD) activity from a Fiber Reinforced Plastic (FRP) material could be triggered by mechanical force by generating initial charges that needed for an electron avalanche. Additionally, the type of partial discharge (PD) in a Fiber Reinforced Plastic (FRP) material may be transited from a Townsend-type to a streamer-type when the size of voids is expanded by mechanical stress, which leads to the increase of partial discharge (PD) magnitude. In [52], partial discharge (PD) activities are observed in Fiber Reinforced Plastic (FRP) composites when exposed to mechanical stress. While for those that are not exposed to mechanical stress, no single partial discharge (PD) is observed. All of these researches point out that the interface between the fiber and the polymer resin in a Fiber Reinforced Plastic (FRP) composite is critical, where micro cracks/voids are inevitable to form. These micro cracks/voids may be elongated, widen and narrowed down by applied mechanical stress and resultantly lead to variation of partial discharge (PD) activities.

2.3.3 Fiber Reinforced Plastic (FRP) Properties in Consideration

The Fiber Reinforced Plastic (FRP) core material's behavior is hard to predict through theoretical analysis. For instance, even though the electric field magnitude calculated numerically in the Fiber Reinforced Plastic (FRP) core with operational conditions is lower than the acceptable value, whether internal partial discharge (PD) happens or not highly depends on imperfections in the core material. Experiments are necessary to investigate the material's electrical properties considering operational conditions.

2.3.3.1 Internal Partial Discharge (PD) Performance

As discussed before, internal partial discharge (PD) is critical for material aging and degradation in long-term operation. Thus, the internal partial discharge (PD) performance in the Fiber Reinforced Plastic (FRP) core with power frequency voltage is of interest. Partial discharge (PD) characteristics, including partial discharge (PD) inception voltage, partial discharge (PD) levels, partial discharge (PD) repetitive rate and partial discharge (PD) patterns with the voltage phase angle are to be measured, because they reflect the imperfection conditions in the solid insulation material.

2.3.3.2 Relative Permittivity

According to IEC Standard 60250 [53], relative permittivity ε_r of a dielectric material is defined as

$$\varepsilon_r = \frac{C_x}{C_0} \tag{2.1}$$

where C_x is the capacitance of a capacitor C, whose electrodes are separated and surrounded by the dielectric material in question; C_0 is the capacitance of the same capacitor C, however dielectric material in it is replaced by vacuum.

ε_r is an important parameter referring to the dielectric polarization of a dielectric material when exposed to electrical stress. Within a combination insulation with different dielectric materials, electric field magnitude in the dielectric with a lower ε_r tends to be higher than that with a higher ε_r. In the case of the composite cross-arm, there is a combination insulation comprised by Fiber Reinforced Plastic (FRP) material, air and weather sheds. Thus, the determination of ε_r of the Fiber Reinforced Plastic (FRP) core is critical, which affects electric field distribution in the cross-arm. Also, ε_r of the Fiber Reinforced Plastic (FRP) core is indispensable for electric field numerical calculation by Finite Element Model (FEM).

2.3.3.3 Dielectric Dissipation Factor and Capacitance

All kinds of dielectrics can be characterized by a capacitance and a magnitude of power dissipation (also named as power loss) [54]. Figure 2.8 shows the equivalent circuit of a dielectric material. The dielectric material in question is always able to be modeled in two kinds of equivalent circuits, in series or in parallel. C_s or C_p is the equivalent capacitance of the dielectric material, while R_s or R_p represents the power dissipation, or power loss, when the dielectric is exposed to electrical stress.

The power loss is comprised of dielectric polarization loss and electrical conductivity loss if the dielectric material is a new material and in good condition. Thus the measurement of power loss is necessary for material quality control, since a too high power loss is undesirable, as it may lead to thermal breakdown in the worst case.

When the material experiences deterioration and aging, or contains imperfections which lead to partial discharge (PD), the power loss will increase. Thus, another advantage of knowing the power loss is to predict whether aging or partial discharge (PD) happens in the dielectric material in question.

In order to present the power loss, the dielectric dissipation factor, $\tan\delta$, is introduced, which is defined as ratio of active power to reactive power in the dielectric. Thus, according to the equivalent circuits of the dielectric in Fig. 2.8, $\tan\delta$ can be calculated by

$$\begin{aligned} \tan\delta &= \omega C_s R_s \\ \tan\delta &= \frac{1}{\omega C_p R_p} \end{aligned} \quad . \tag{2.2}$$

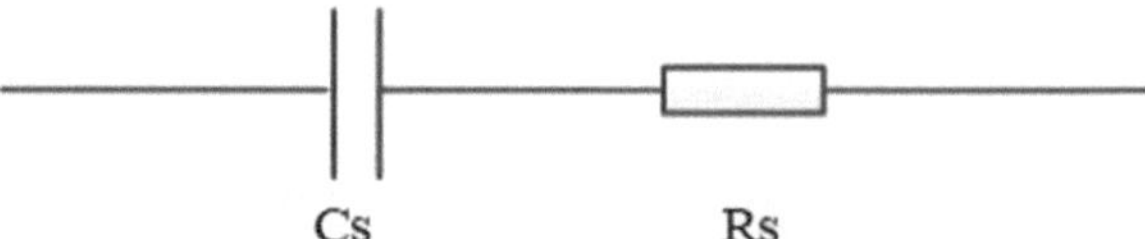

(a) Equivalent series circuit

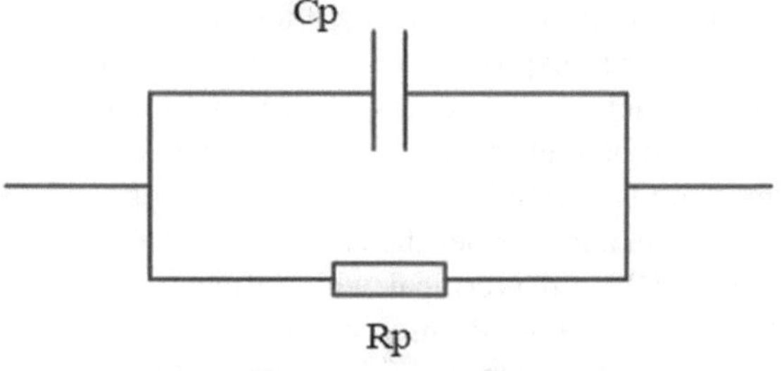

(b) Equivalent parallel circuit

Fig. 2.8 Equivalent circuit to model a dielectric material [54]

2.4 Electrical Test on Fiber Reinforced Plastic (FRP) Composites

In order to measure the electrical properties of the Fiber Reinforced Plastic (FRP) core material, a AC test setup was built up to provide power frequency voltage in the HV lab at the Department of Energy Technology, Aalborg University. Two measurement circuits for partial discharge (PD) measuring and dielectric properties measuring were also built up.

2.4.1 Test Circuit and Setup

2.4.1.1 AC Test Circuit and Setup

The AC test circuit is shown in Fig. 2.9. T represents the transformer (220/100,000 V, 5 kVA), manufactured by HIGHVOLT Prüftechnik Dresden GmbH. R_0 is a protective resistor with a resistance of 50 kΩ. ES is an operational switch.

C_1 with the capacitance of 100 pF is the high voltage capacitor in the capacitive voltage divider comprised of C_1 and C_2, which takes the function of measuring applied voltage to the test object. The SM615 instrument is a peak value detecting instrument with three built-in low voltage capacitors, i.e three values of C_2, thus three scale ratios-$1:1$, $1:2$ and $1:5$ can be achieved. Correspondingly, the full scale measuring values of 50, 100 and 200 kV are available. A breakdown detecting device, represented by BD in Fig. 2.9, is installed between C_1 and SM615, in case overvoltage is presented.

After the capacitive voltage divider, the test object S is connected. Figure 2.10 shows the electrodes arrangement between which the test object in question is placed. The HV electrode is a circular plate with a diameter of 110 mm, while the LV electrode is circular plate with a diameter of 100 mm. The test object is placed between the HV and LV electrode. There is an annular electrode surrounding the LV electrode, which is called 'Guard electrode'. The aim of the guard electrode is to eliminate surface leakage current component from the output signal, making sure the measured current is flowing though the test object.

In some cases, such as in partial discharge (PD) measurement test, the electrodes with the test object must be put into an insulating oil tank to eliminate the effect of electrical discharge (corona discharge) in the air. Figure 2.11 shows the arrangement (the tank is not shown in the figure). The oil meets the special requirements for transformer oil defined in IEC Standard 60296 [55]. The electrical strength of the oil is more than 70 kV/mm.

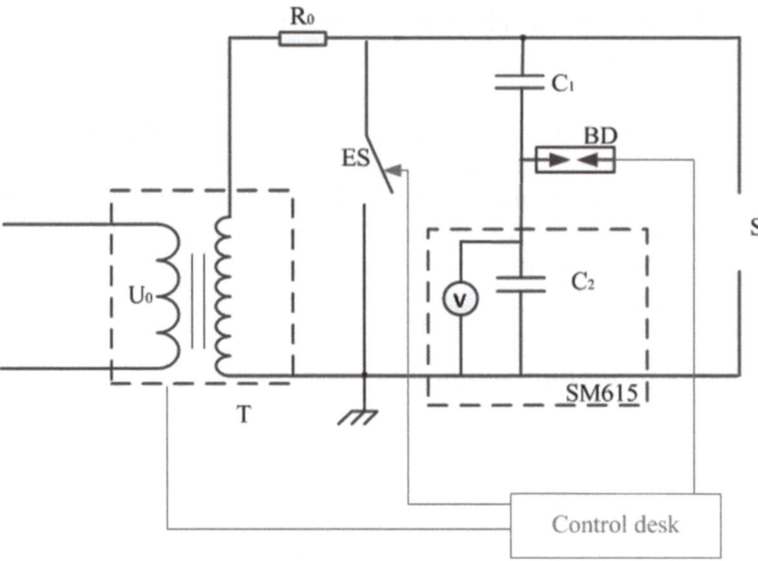

Fig. 2.9 Schematic of AC test circuit

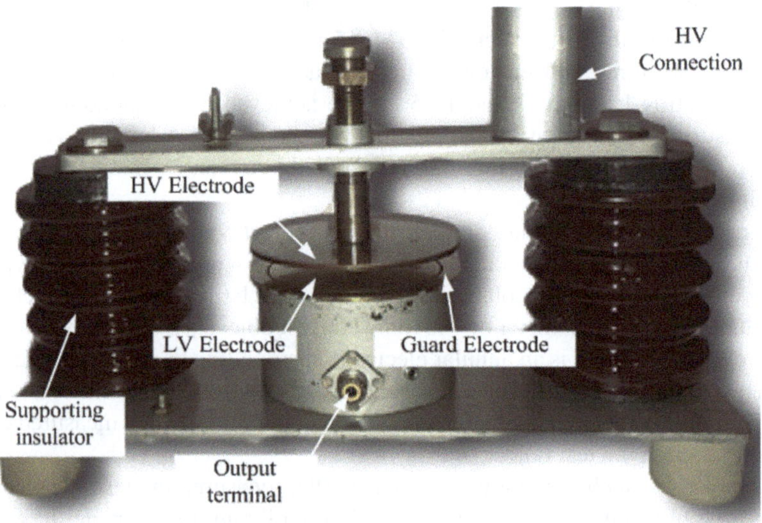

Fig. 2.10 Electrodes arrangement

Fig. 2.11 Electrodes with test object in an insulating oil tank

2.4.1.2 Partial Discharge (PD) Measurement Circuit and Setup

Partial Discharge (PD) Measurement Mechanism

The three-capacitor model is widely used when analyzing partial discharge (PD) activities in a cavity inside of a solid insulation. Figure 2.12 shows the model.

In Fig. 2.12a, the solid insulation material with a single cavity is placed between two parallel plate electrodes and exposed to an electrical stress. C_c represents the equivalent capacitance of the cavity. C_b' and C_b'' represent the equivalent capacitance of the insulation material that is series connected with the cavity. C_a' and C_a'' represent the equivalent capacitance of the rest 'healthy' insulation material.

Compared circuits in Fig. 2.12a and b, the following equation exists:

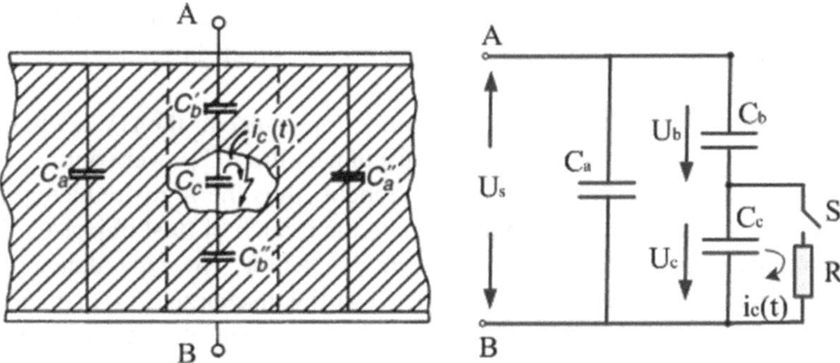

Fig. 2.12 Partial discharge (PD) model in the cavity of a solid insulation: **a** Schematic of a solid insulation comprising a cavity; **b** equivalent circuit [54]

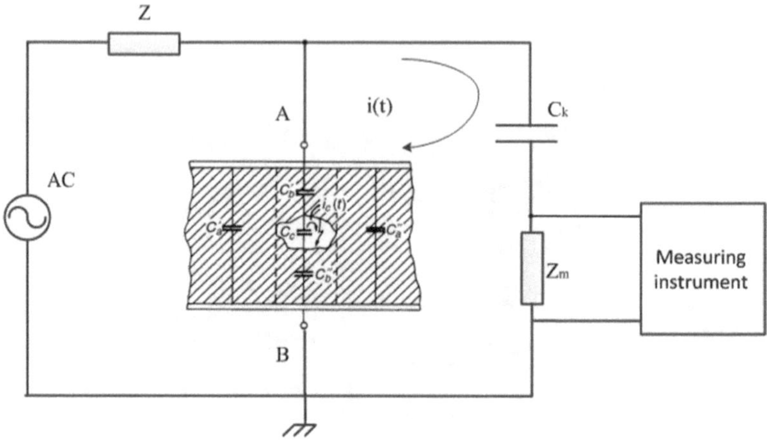

Fig. 2.13 Partial discharge (PD) test circuit

$$\begin{cases} C_a = C_a' + C_a'' \\ C_b = \frac{C_b' C_b''}{C_b' + C_b''} \end{cases} \tag{2.3}$$

When the magnitude of the electric field at the cavity exceeds the critical value, electrical discharge happens in the cavity, which is presented by switching on S in Fig. 2.12b. As a result, C_c discharges through R and there is a current impulse flowing through the branch C_c-S-R. ΔU_c is the voltage drop at the cavity when PD happens and the charge magnitude Δq_c involved in the cavity when partial discharge (PD) happens can be calculated by

$$\Delta q_c = \Delta U_c * C_c = \int i_c(t)dt \tag{2.4}$$

However, values of ΔU_c, C_c and $i_c(t)$ are neither known nor able to measure. In order to get information of the partial discharge (PD) activity, the partial discharge (PD) measuring circuit as Fig. 2.13 is established, which is recommend by IEC Standard 60270 [47].

Z is a low-pass filter, comprising either only impedance of the HV lead or a well designed inductor. The function of Z is to disconnect the AC power source from the main circuit when partial discharge (PD) happens, thus the high-frequency partial discharge (PD) current pulse $i(t)$ only flows in the branch of 'test object-C_k-Z_m'.

C_k is a partial discharge (PD)-free coupling capacitor. The capacitance of C_k is much larger than that of the test object. When a partial discharge (PD) happens in the test object, C_k is regarded as a power source, which compensates the voltage drop at the test object.

In the test object, the three capacitors fulfill the following relations:

$$\begin{cases} C_a \gg C_b \\ C_c \gg C_b \end{cases} \tag{2.5}$$

Thus, the equivalent capacitance C_t of the whole test object can be approximated as

$$C_t = C_a + \frac{C_b C_c}{C_b + C_c} \approx C_a + C_b \tag{2.6}$$

Since C_k is regarded as a power source when PD happens, the charge deviation Δq at the test object when partial discharge (PD) happens can be calculated as

$$\Delta q = C_t * \Delta U_a \approx (C_a + C_b)\Delta U_a \tag{2.7}$$

From Fig. 2.12b, Eq. (2.8) is obtained

$$\Delta U_a = \frac{C_b}{C_a + C_b} * \Delta U_c \tag{2.8}$$

Substituting Eq. (2.8) into Eq. (2.7), Δq is calculated by

$$\Delta q = C_b * \Delta U_c \tag{2.9}$$

Equation (2.10) indicates that the value of charge variation at the test object does not equal to that of the real charge involved in the cavity when partial discharge (PD) happens.

$$\Delta q_c \gg \Delta q \tag{2.10}$$

However, it is obvious that Δq is proportional to U_c, considering the value of C_b increases with the increase of the cavity dimension. Thus, Δq contains information of the partial discharge (PD) activity occurring inside the cavity. Δq is called the **Apparent charge** and can be calculated by

$$\Delta q = \int i(t)dt \tag{2.11}$$

Z_m is a measuring impedance, which transfers the partial discharge (PD) current impulse into voltage signals and measured by the following measuring instrument. Actually, Z_m is a band-pass filter, which has a lower and upper limit frequency f_1 and f_2. It can be approximated that only frequency components of the partial discharge (PD) signal within f_1 and f_2 can be measured. Thus, the determination of f_1 and f_2 as well as the band width Δf is critical for the accuracy of partial discharge (PD) measurement. IEC Standard 60250 recommends values of f_1, f_2 and Δf as follows [47].

For a wide-band filter:

$$\begin{gathered} 30\ \text{kHz} \leq f_1 \leq 100\ \text{kHz} \\ f_2 \leq 500\ \text{kHz} \\ 100\ \text{kHz} \leq \Delta f \leq 400\ \text{kHz} \end{gathered} \tag{2.12}$$

For a narrow-band filter:

$$\begin{gathered} 9\ \text{kHz} \leq \Delta f \leq 30\ \text{kHz} \\ 50\ \text{kHz} \leq \frac{f_1+f_2}{2} \leq 1\ \text{MHz} \end{gathered} \tag{2.13}$$

From Eqs. (2.4)–(2.11), the conclusion can be drawn that the partial discharge (PD) activities occurring in cavities within the solid insulation material, which are unable to measure directly, can be analyzed and evaluated by measuring the apparent charge Δq with the measuring circuit in Fig. 2.13.

Before measuring the apparent charge, calibration must be done. The aim of the partial discharge (PD) calibration is to match the charge reading of the partial discharge (PD) measurement instrument with the real value of the apparent charge Δq. The process of partial discharge (PD) calibration contains following steps:

1. Disconnect the circuit with the power source;
2. Inject a certain amount of charge Δq_0 into the test object with a fixed frequency and get the corresponding reading M_0 of the measuring instrument;
3. Calculate the scale factor S_f by Eq. (2.14)

$$S_f = \frac{\Delta q_0}{M_0} \tag{2.14}$$

Thus, in the partial discharge (PD) test, the value of apparent charge Δq_x in the test object can be calculated with the reading M_1 of the measuring instrument by

$$\Delta q_x = M_1 * S_f \tag{2.15}$$

Partial Discharge (PD) Test Circuit and Setup

Figure 2.14 indicates the partial discharge (PD) test circuit applied in the present thesis.

C_x represents the equivalent capacitance of the test object in question. C_k is a coupling capacitor, with a capacitance of 1 nF. The maximum voltage applied to C_k is 100 kV_{rms} with which partial discharge (PD) level from C_k is less than 1 pC.

Z_m is the measuring impedance manufactured by OMICRON. The maximum current permitted through Z_m is 2 A and it contains two built-in low-arm capacitors with capacitance of 30 μF and 120 μF. The partial discharge (PD) signal output frequency range of Z_m is from 20 kHz to 6 MHz. Meanwhile, C_k and the low-arm

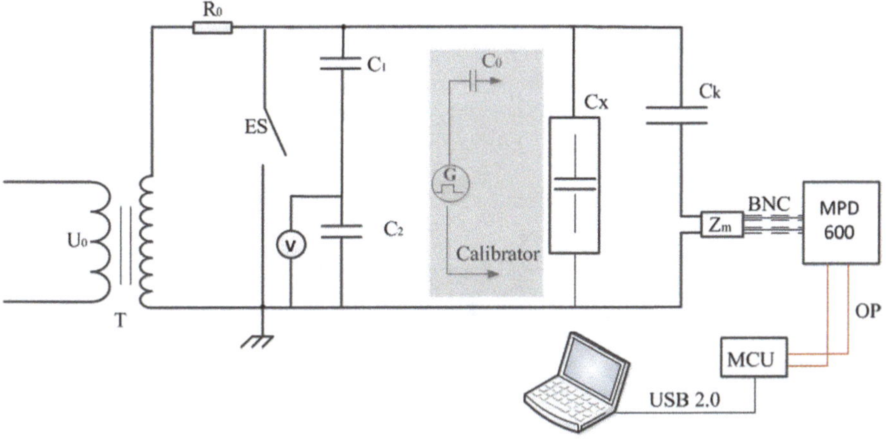

Fig. 2.14 Schematic of partial discharge (PD) test circuit

capacitor in Z_m comprise a capacitive voltage divider and thus the voltage magnitude and phase at the test object can be obtained.

MPD 600 is the partial discharge (PD) signal measurement unit, manufactured by OMICRON. The input partial discharge (PD) signal frequency range is from 0 kHz to 20 MHz, while the input voltage frequency range is from 0.1 Hz to 2.1 kHz. The maximum input voltage for partial discharge (PD) signals is 10 V_{rms} while it is 60 V_{rms} for input voltage signals. There is a built-in protection circuity inside to protect it from over-currents and over-voltages. MCU is a control unit between the measuring software and the MPD 600 unit.

Z_m is connected through two BNC cables with the MPD 600 unit. While MPD 600 is connected with MCU by two optical fiber cables, thus they are totally electrical insulated. The electrical insulation between partial discharge (PD) measuring unit and the control unit makes the system more secure, especially in the case that the test object is not grounded. The BNC cables are kept as short as possible during test, in order to minimize the inductance of the cable.

The whole MPD system is controlled by an integrated MPD/MI software. It enables real-time measurement of partial discharge (PD) signals and analysis.

The calibrator, in fact, is a charge generator, which outputs a certain value of charge with a fixed frequency of 300 Hz. The rise time of the charge pulse is as low as 10 ns. The values of the output charge are ±10, ±20, ±50 and ±100 pC.

2.4.1.3 Capacitance and Dissipation Factor Test and Setup

Figure 2.15 indicates the schematic of the test circuit for the measurement of capacitance C_x and dielectric dissipation factor $\tan\delta$ of the test object.

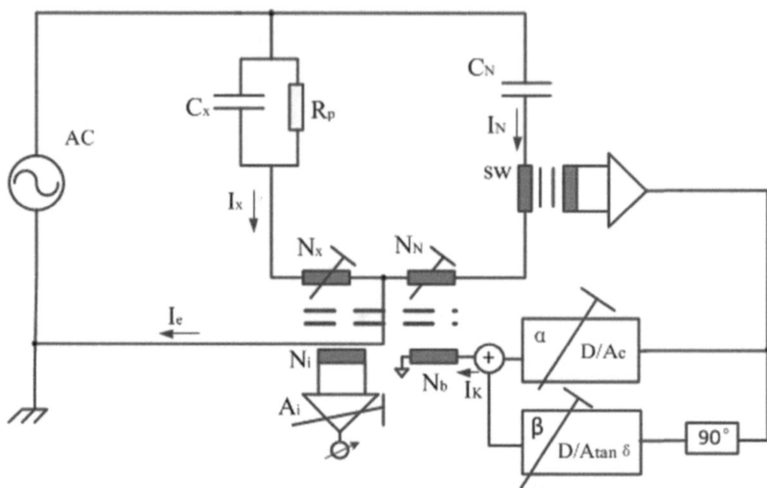

Fig. 2.15 Schematic of automatic high voltage current comparator bridge [56]

The circuit in Fig. 2.15 is called 'current comparator bridge'. C_x and R_p represent the equivalent capacitance and resistance of the test object as Fig. 2.8b shows. C_N is a nitrogen-filled standard capacitor with a capacitance of 37.2 pF. The power loss of C_N is very low, thus it can be neglected in the test.

The aim of the current bridge is to compare the current flowing through the test object and through C_N, which are presented by I_X and I_N respectively. SW is comprised of a current transformer and a current-voltage converter which takes part of I_N and transfers it into voltage. D/A_c and $D/A_{\tan\delta}$ is an analog-digital-converter in phase and 90° phase lag respectively. Through D/A_c and $D/A_{\tan\delta}$, part of I_N is fed into winding N_b.

N_X, N_N and N_b are windings winded at the same iron core of high permeability with the shape of a ring. If the product of ampere and windings is zero, the magnetic flux in the iron core is zero. The voltage at winding N_i is proportional to the magnetic flux in the iron core. The very small voltage at N_i is amplified by the amplifier A_i and detected by a microprocessor.

When the parameters of N_X, N_N, N_b, D/A_c and $D/A_{\tan\delta}$ are adjusted properly, the magnetic flux in the iron core will be zero thus the microprocessor will detect a zero voltage. This state is called 'balance state'. With a balanced state, the values of C_X and R_p can be calculated with the parameters in the system as following.

The current I_X through the test object is:

$$I_X = U_t * (\frac{1}{R_p} + j\omega C_X) \tag{2.16}$$

where U_t is the applied test voltage.

The current I_N through the standard capacitor is:

$$I_N = U_t * j\omega C_N \tag{2.17}$$

For balanced state:

$$U_t * N_x * (j\omega C_X + \frac{1}{R_p}) = U_t * j\omega C_N * \left[N_n + k * (\alpha - j\beta) * N_b'\right] \tag{2.18}$$

where N_x, N_n and N_b' are number of windings of N_X, N_N and N_b respectively. α, β and k are parameters for the analog-digital-converters D/A_C and $D/A_{\tan\delta}$.

Thus, the equivalent capacitance C_X and the dissipation factor $\tan\delta$ can be calculated by applying the balanced state and Eq. (2.18):

$$C_X = C_N * \frac{N_n + k * N_b' * \alpha}{N_x} \tag{2.19}$$

$$\tan\delta = \frac{N_b'}{N_n + k * N_b' * \alpha} * \beta \tag{2.20}$$

The automatic high voltage current comparator bridge applied in the present thesis is manufactured by PRESCO AG with the type of TG-3MOD.

The maximum current permissable through the test object and the standard capacitor are 5 A and 30 mA respectively. The uncertainty of capacitance, phase and voltage measurement are $\pm 1\%$, $\pm 0.02\%$ mrad $\pm 1\%$ displayed value and $\pm 1\%$ of range ± 2 digits.

2.4.1.4 Relative Permittivity Test Circuit and Setup

As Eq. (2.1) indicates, the relative permittivity can be obtained if the equivalent capacitance C_x of the test object and the capacitance C_0 are known. C_0 is the capacitance of a capacitor with the same electrode arrangement with the test object, but the electrodes are separated and surrounded by vacuum instead of the dielectric material.

In order to find the ε_r of the test object, the automatic high voltage current comparator bridge in Fig. 2.15 and the electrodes in Fig. 2.10 are applied in the present thesis.

Firstly, place the test object between the electrodes and measure its capacitance C_x by the current comparator bridge. Secondly, remove the test object and measure the capacitance C_0 of the electrodes without the test object. There is an important assumption that the measured C_0 when the electrodes are separated and surrounded by air is approximated equivalent to that when the electrodes are separated and surrounded by vacuum. Finally, the relative permittivity of the test object is calculated by Eq. (2.1) with measured values of C_x and C_0.

2.4.2 *Electrical Test*

In this section, the electrical tests conducted on the Fiber Reinforced Plastic (FRP) core material of the composite cross-arm are introduced and test results are analyzed.

2.4.2.1 Test Sample

Test samples tested in the present thesis are produced by TUCO MARINE Aps.

As introduced in Sect. 2.1.1, in electrical power system application, E-glass fiber is widely used due to its good electrical/mechanical properties as well as low cost. Meanwhile, vinyl ester and epoxy, both of which have good mechanical strength are regarded as promising polymers for the core material of the composite cross-arm. Thus, E-glass reinforced vinyl ester (abbreviated as glass/vinyl ester) and E-glass reinforced epoxy (abbreviated as glass/epoxy) are chosen for testing.

Manufacturing methods have great effects on the Fiber Reinforced Plastic (FRP) material's quality. Samples manufactured by three methods, including hand lay-up moulding, vacuum assisted resin transfer moulding and filament winding are chosen.

Dimension of the material sample also has effects on the final electrical performance, the samples used in the tests are square plates with length and width of 150 mm, while the thickness of the plate are 1–10 mm.

Figure 2.16 shows typical samples plates-glass/vinyl ester manufactured by filament winding.

Fig. 2.16 Photo of glass/vinyl ester sample plates manufactured by filament winding with dimensions: $L = 150$, $M = 150$ and $T = 10$ mm

2.4.2.2 Partial Discharge (PD) Test

The partial discharge (PD) test was performed to investigate the partial discharge (PD) performance of the Fiber Reinforced Plastic (FRP) samples with power frequency voltage.

As introduced in Sect. 2.3.2.2, the maximum electric field magnitude in the core material of the composite cross-arm is 0.6 kV_{rms}/mm, while the acceptable criterion for maximum electric field magnitude in the core material is 3 kV_{rms}/mm. Thus, the partial discharge (PD) inception voltage and inception electric field magnitude of the Fiber Reinforced Plastic (FRP) samples were of our primary interest. The measured values were compared with the calculated value and the criterion.

Also, other partial discharge (PD) characters, such as partial discharge (PD) patterns, average and maximum partial discharge (PD) levels were also measured to assess the Fiber Reinforced Plastic (FRP) materials' electrical quality.

Test Procedure

Before test, all the samples were cleaned by ethyl alcohol to make sure no contamination remained on the surface.

Then the Fiber Reinforced Plastic (FRP) samples were placed between the electrodes and they were put into the oil tank as shown in Fig 2.11. The test circuit was implemented according to the schematic in Fig. 2.14.

The partial discharge (PD) measurement unit was set in accordance with requirements of IEC Standard 60270 [47]. A center frequency $f_m = 250$ kHz and a bandwidth of 300 kHz were chosen for filtering. A value of 2 pC was set as the partial discharge (PD) detective threshold value. Partial discharge (PD) calibration was done. Also, the partial discharge (PD) level from background noise was measured as 0.1 pC approximately.

The voltage was raised from zero with a constant speed until any single partial discharge (PD) activity was detected. The Root-Mean-Square (RMS) value of the applied voltage at which the partial discharge (PD) activities was detected for the first time was recorded as 'partial discharge (PD) inception voltage'. For each kind of material, manufactured by a specific method and with a specific dimension (thickness), a total number of 5 specimens were tested. For each material sample i ($i = 1, 2, 3, 4, 5$), 5 measurements were performed in sequence at an interval of 1 min, i.e. after each measurement of partial discharge (PD) inception voltage, the applied voltage was reduced to zero and after 1 min, the voltage was raised again until another partial discharge (PD) inception voltage was recorded. The 5 partial discharge (PD) inception voltages were recorded as $V_{inc-i-n}$ ($n = 1, 2, 3, 4, 5$) correspondingly. In total, for each kind of material manufactured by a specific method and with a specific thickness, 25 partial discharge (PD) inception voltages were obtained. The arithmetic mean value of $V_{inc-i-n}$ was the final partial discharge (PD) inception voltage of the certain material. The standard deviation σ of the 25 partial discharge (PD) inception voltages was also calculated.

In order to make sure all measured partial discharge (PD) signals were from material samples instead of from the test setup, after partial discharge (PD) measurement on Fiber Reinforced Plastic (FRP) samples, a voltage with Root-Mean-Square (RMS) value of 1.1 times the highest partial discharge (PD) inception voltage of all recordings was applied to the system without any sample and partial discharge (PD) level from the system was measured. It proved that there was no partial discharge (PD) activity recorded because of the test setup.

After V_{inc} was recorded, the critical electric field magnitude, i.e. partial discharge (PD) inception electric field E_{inc} was calculated by

$$E_{inc} = \frac{V_{inc}}{T} \tag{2.21}$$

where T is the thickness of the sample in mm.

For each measurement, the voltage was kept as $1.1 \times V_{inc-n}$ for 60 s, during which period, partial discharge (PD) patterns with the voltage phase (also called PRPD-phase resolved partial discharge patterns), average/maximum partial discharge (PD) magnitude and partial discharge (PD) repetitive rate were recorded.

Partial Discharge (PD) Patterns

Figure 2.17 shows a typical phase resolved partial discharge (PRPD) patterns measured from a glass/epoxy sample with a thickness of 10 mm manufactured by filament winding. The applied voltage V is 1.1 times the partial discharge (PD) inception voltage, i.e. $V = 35\ \text{kV}_{\text{rms}}$.

It can be seen from Fig. 2.17 that internal partial discharge (PD) happens from the sample, which is expected. There are two obvious partial discharge (PD) signals, as the blue and red frame indicate in the figure. It proves that there are at least two partial discharge (PD) sources from the sample.

In order to corresponds the partial discharge (PD) activity with the quality of the sample, an optical observation was performed at the sample. An optical microscope with a tenfold magnification is applied to observe the internal condition of the sample with a background light source against the sample. Figure 2.18 shows that there is an observable air bubble from the sample with a diameter of 1 mm approximately. This air bubble is a possible internal partial discharge (PD) source at $V = 3.85\ \text{kV}_{\text{rms}}$. It should be noted that there may be other defects in the sample, since lights going to the sample will be scattered by numerous fibers and the sample cannot be looked through.

Partial Discharge (PD) Inception Electric Field

Figure 2.19 shows Root-Mean-Square (RMS) values of electric field at which partial discharge (PD) activities were measured from Fiber Reinforced Plastic (FRP) samples manufactured by different methods. The error bar in Fig. 2.19 indicates the standard deviation σ of the measurement data.

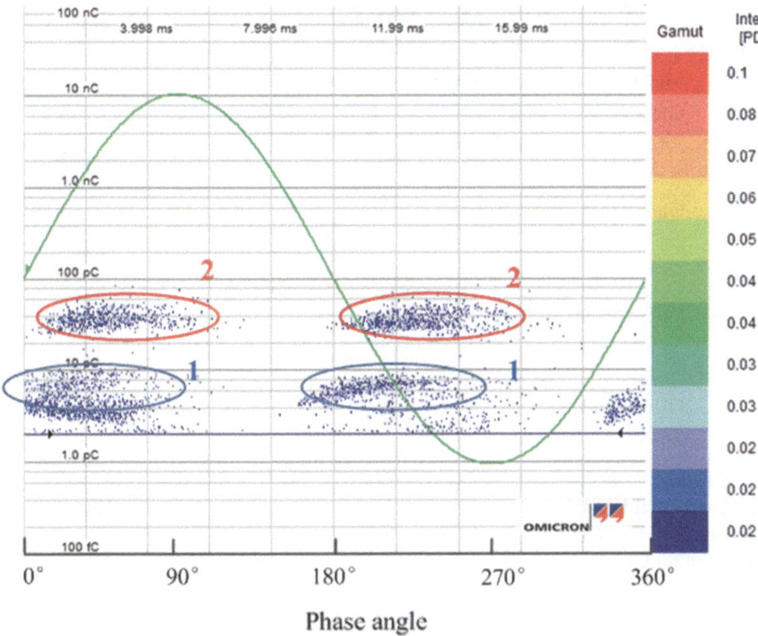

Fig. 2.17 A typical phase resolved partial discharge (PRPD) patterns from a glass/epoxy sample with a thickness of 10 mm manufactured by filament winding at $V = 35\ \mathrm{kV_{rms}}$

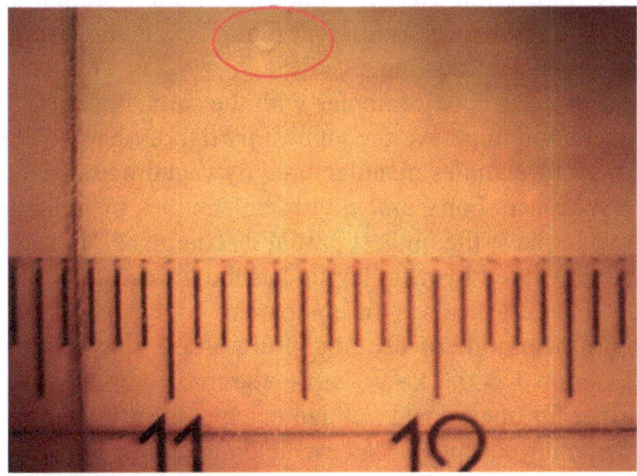

Fig. 2.18 Optical observation of the same glass/epoxy sample in Fig. 2.17

Table 2.1 Comparison of partial discharge (PD) inception electric field magnitude for all Fiber Reinforced Plastic (FRP) samples with two criteria

Thickness	Material	Manufacturing method	Criterion 1	Criterion 2
1 mm	Glass/epoxy	1#	√	×
		2#	√	√
		3#	√	√
	Glass/vinyl ester	1#	√	×
		2#	√	√
		3#	√	√
10 mm	Glass/epoxy	1#	×	×
		2#	√	×
		3#	√	√
	Glass/vinyl ester	1#	×	×
		2#	√	×
		3#	√	×

1#: Hand lay-up moulding method
2#: Vacuum assisted resin transfer moulding method
3#: Filament winding method

For all samples, it can be seen from Fig. 2.19 that values of inception electric field for thinner ones were bigger than that for thicker ones. It proves that the thickness of the sample has effects on the characteristics of defects in a Fiber Reinforced Plastic (FRP) material. Thicker samples contain more laminates. It is possible that more voids grow between different laminates, leading to a lower partial discharge (PD) inception electric field.

For glass/epoxy samples with the same thickness, samples manufactured by filament winding have the highest partial discharge (PD) inception electric field, followed by samples manufactured by vacuum assisted resin transfer and hand lay-up in sequence. For glass/vinyl ester, samples manufactured by vacuum assisted resin transfer have the highest partial discharge (PD) inception electric field and the one manufactured by hand lay-up has the lowest.

Table 2.1 displays the comparison of the partial discharge (PD) inception electric field magnitude of all Fiber Reinforced Plastic (FRP) samples with two criteria. Criterion 1 is 0.6 kV_{rms}/mm—the maximum electric field in the composite cross-arm with operational voltage calculated in [57]. Criterion 2 is 3 kV_{rms}/mm—the maximum acceptable electric field magnitude in Fiber Reinforced Plastic (FRP) core in composite insulator/cross-arm which is widely used in industry. If the measured E_{inc} is higher than the criterion, a '√' is marked, otherwise a '×' is marked.

It is obvious from Table 2.1 that only glass/epoxy samples manufactured by filament winding method fulfills two criteria simultaneously with different thicknesses.

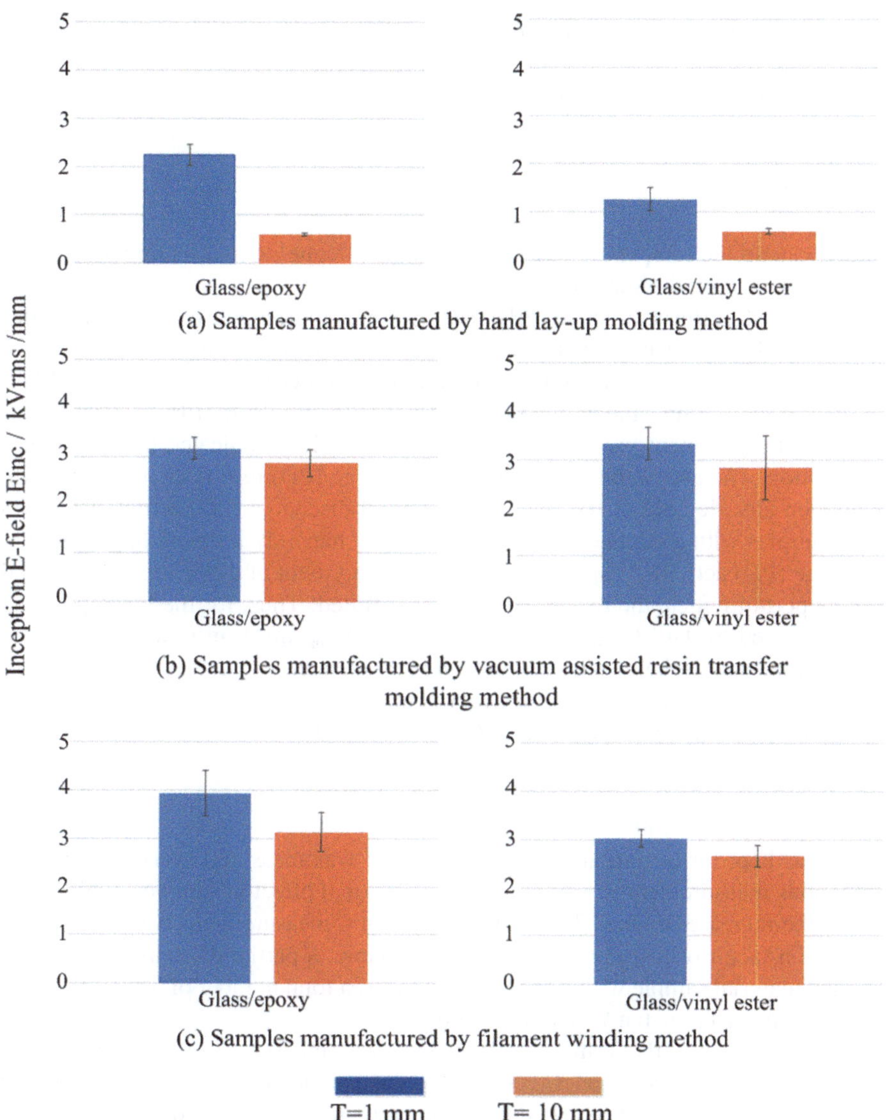

Fig. 2.19 Root-Mean-Square (RMS) values of partial discharge (PD) inception electric field for samples manufactured by different methods

2.4.2.3 Dielectric Characteristics Test

Dielectric characteristics, including dissipation factors $\tan \delta$ and relative permittivities ε_r were tested.

Test Procedure

Before test, all the samples were cleaned by ethyl alcohol to make sure no contamination remained on the surface.

The Fiber Reinforced Plastic (FRP) samples were placed between the electrodes as shown in Fig. 2.10 and the circuit in Fig. 2.15 was established.

As the maximum electric field magnitude in the Fiber Reinforced Plastic (FRP) core is 0.6 $\mathrm{kV_{rms}/mm}$, the applied electric field magnitude E_0 for dielectric characteristic measurement was set as 0.6 $\mathrm{kV_{rms}/mm}$, which indicated the worst case in operation. The above-mentioned partial discharge (PD) results display that, with an applied electric field of 0.6 $\mathrm{kV_{rms}/mm}$, partial discharge (PD) activities happen in all samples with a thickness of 10 mm made by hand lay-up moulding. If partial discharge (PD) activities happen during the measurement of dissipation factors and relative permittivities, the results will be exaggerated. Thus, for these samples, E_0 was decreased to 0.48 $\mathrm{kV_{rms}/mm}$ (80% of 0.6 $\mathrm{kV_{rms}/mm}$), in order to eliminate effects of partial discharge (PD) signals on measurement.

The voltage was raised slowly from zero but with a constant speed until the target voltage U_0 was reached. The value of U_0 is calculated by

$$U_0 = E_0 \times T \tag{2.22}$$

U_0 was kept for 60 min and the value of $\tan \delta$ was measured every 10 min. The continuous measurement was to consider the thermal effect in samples, which may lead to the increase of $\tan \delta$. The average value of 6 measurements was recorded as dissipation factor of a specific sample. As the same in partial discharge (PD) tests, for each material sample with a specific thickness, a total number of 5 samples were tested and the average $\tan \delta$ was the final result.

For each sample, the value of capacitance was measured for 5 times and the average value was recored as C_x. Then, the capacitance of the electrodes without samples was measured. The capacitance when there is a 10 mm distance between two electrodes was recored as C_{0-10}, while the value was recored as C_{0-1} with a 1 mm distance. Again, each value was measured for 5 times and the average value was recorded, to reduce measuring error. Then the relative permittivity of a specific sample was calculated according to Eq. (2.1).

It should be noted that, all the measurement was performed with a power frequency voltage and a temperature of 20 °C.

Test Results of tan δ

Figure 2.20 displays values of tan δ for Fiber Reinforced Plastic (FRP) samples manufactured by different methods. The standard deviations are also shown by error bars in the figure.

It can be seen from Fig. 2.20 that there is a slight difference in values of tan δ for thinner samples and thicker ones, i.e. the former is smaller than the latter.

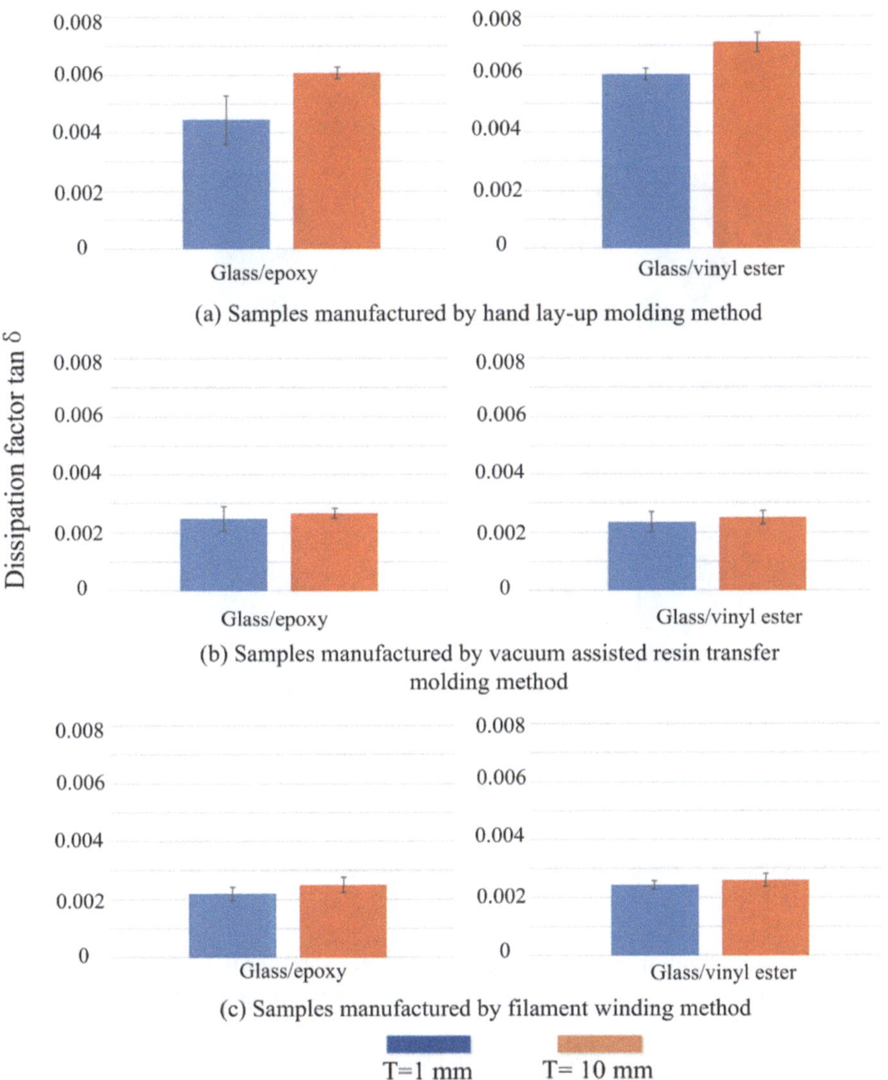

Fig. 2.20 Values of tan δ for samples manufactured by different methods

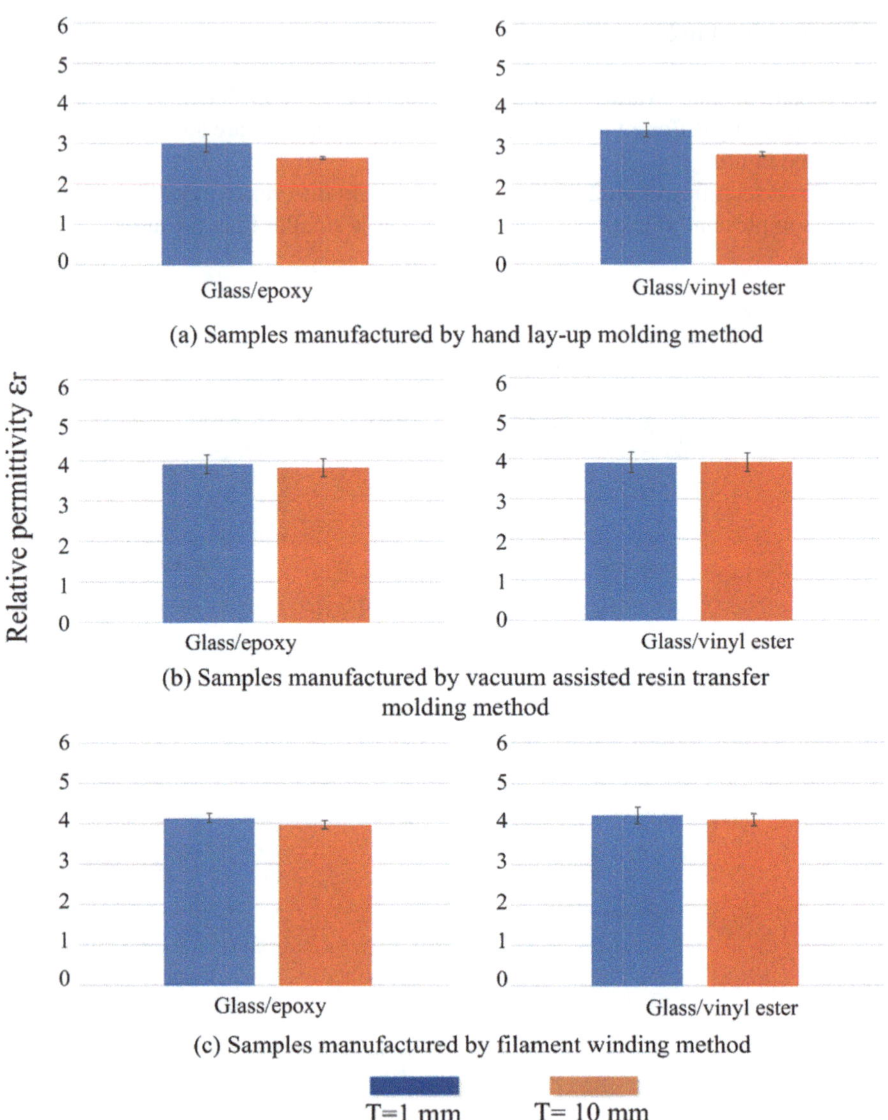

Fig. 2.21 Values of ε_r for samples manufactured by different methods

For the same kind of material, values of tan δ for samples manufactured by hand lay-up, vacuum assisted resin transfer and filament winding are in the order from large to small.

Comparing glass/epoxy and glass/vinyl ester samples manufactured by the same method with the same thickness, values of tan δ do not differ a lot, except for samples manufactured by hand lay-up moulding method. The reason for that is still unclear.

There is not any international criterion for the value of tan δ of Fiber Reinforced Plastic (FRP) materials as core material of composite insulators/cross-arms. In a relative reference [22], which also applies Fiber Reinforced Plastic (FRP) material to composite transmission towers for 110 kV in China, tan $\delta \leq 0.002$ is set as the criterion. While in a technical report of hollow composite composite insulators for 72–1200 kV by ABB, the value of tan δ is close to 0.005. However, since the dissipation factor reflects the power loss during operation, a low value is always preferred. Among all samples, the glass/epoxy manufactured by filament winding has the lowest value and is more desirable.

Test Results of ε_r

Figure 2.21 displays values of ε_r for Fiber Reinforced Plastic (FRP) samples manufactured by different methods. The standard deviations are also shown by error bars in the figure.

For samples manufactured by hand lay-up moulding method, the value of ε_r for thinner samples is larger than that of thicker ones. While for other samples manufactured by vacuum assisted resin transfer moulding and filament winding method, thickness of the sample does not have obvious effects on the value of ε_r.

For both samples manufactured by vacuum assisted resin transfer moulding and filament winding method, the value of ε_r equals to 4 approximately, which is larger than that of samples made by hand lay-up moulding method.

There is not any recommended value for ε_r of Fiber Reinforced Plastic (FRP) core material in the application to composite insulators/cross-arms. In [22], the glass/epoxy used for the composite material based tower has a relative permittivity of 5 approximately in dry conditions. While in [58], the value is close to 6. Actually, a larger value of ε_r can help to reduce the electric field magnitude in the Fiber Reinforced Plastic (FRP) core. From this point of view, a larger ε_r value is more preferable.

Detailed values of tan δ and ε_r for all samples can be referred to Appendix A.

2.4.3 Discussion

2.4.3.1 Partial Discharge (PD) Performance

From partial discharge (PD) tests, glass/epoxy samples manufactured by filament winding have the highest partial discharge (PD) inception electric field magnitude for a certain thickness, while samples manufactured by hand lay-up method have the lowest value.

The above-mentioned phenomenon is possible to be explained by the difference in fiber volume fraction of different samples. The fiber volume fraction is defined as the ratio of the fiber volume to the total volume of the bulk material. In [59], effects

Table 2.2 Fiber volume fraction dependent on manufacturing method [59]

Manufacturing method	Fiber volume fraction (%)
Hand lay-up	30
Compression moulding	40
Filament winding	60–85
Vacuum moulding (bagging)	50–80

of different manufacturing methods on the fiber volume fraction in Fiber Reinforced Plastic (FRP) materials have been investigated, as shown in Table 2.2.

Table 2.2 shows that samples manufactured by filament winding have the highest fiber volume fraction. The hand lay-up method leads to the lowest fiber volume fraction. The fiber volume fraction in samples manufactured by vacuum assisted resin transfer method should be similar with that of samples manufactured by the vacuum moulding (bagging) method in Table 2.2.

In [60], effects of fiber volume fraction on voids' characteristics in glass/epoxy samples are studied, and it proves that the increase of fiber content in the bulk material leads to a moderate decrease of void number. Meanwhile, with a higher fiber content, the formation of irregularly-shaped voids- neither spheric nor ellipsoidal- is reduced.

As a result, the number of voids in samples manufactured by filament winding should be the smallest, compared with samples manufactured by the other two methods. On the other hand, the number of voids in samples manufactured by the hand lay-up method is the largest. As less voids indicate less internal partial discharge (PD) sources, the partial discharge (PD) inception electric field of samples manufactured by filament winding is larger.

Additionally, electric field concentration is intensified in irregularly-shaped voids in a bulk material. Thus, partial discharge (PD) inception voltage is higher in samples manufactured by filament winding, in which less irregularly-shaped voids are found.

Another reason that samples manufactured by hand lay-up have the lowest partial discharge (PD) inception electric field lies in the intrinsic shortcoming of the method. As described in Sect. 2.1.3, during a hand lay-up moulding process, the fiber fabric is pressed by a roller and rabbler to force the liquid resin infiltrate in to it. It is obvious that the infiltration is not through without help of extra pressure. As a result, voids are easy to form between fibers and the resin. Additionally, entrapped gases in the composite cannot be exhausted throughly by only squeezing the composites by the roller or rabbler.

It is worth noting that there are many other parameters which have effects on voids' formation in a Fiber Reinforced Plastic (FRP) material. For example, the viscosity of the resin is one of the critical parameters, as a higher viscosity always causes more voids [61]. The resin viscosity is not only dependent on intrinsic property of the resin, but also affected by other factors, such as solvents added in the fiber impregnation process and applied temperature over time during the curing process [61]. The resin flow is also an important factor in the case of liquid resin flowing though the fiber fabrics and infiltrating, such as in vacuum assisted resin transfer moulding process.

The resin flow have impacts on the size and location of voids formed in the final Fiber Reinforced Plastic (FRP) material. And the resin flow is dependent on the fiber fabric's geometry, moulding complexity and resin intrinsic property [62]. Applied compaction pressure is critical which determine whether as much as entrapped gases can escape from the final bulk material [63]. Additionally, temperature cycle during curing process is important that affects the void number, shape and spatial distribution in the bulk material [64].

In a word, the voids characteristics in a Fiber Reinforced Plastic (FRP) materials, such as number, size, location, shape, etc., are results caused by numbers of factors. Thus, the internal partial discharge (PD) characteristics, most importantly partial discharge (PD) inception electric field, cannot be predicted accurately beforehand. Partial discharge (PD) measurement is an effective way to understand the internal conditions of a specif Fiber Reinforced Plastic (FRP) material.

As partial discharge (PD) test results show, glass/epoxy samples manufactured by filament winding method have the highest partial discharge (PD) inception electric field and fulfill the two criteria, i.e. the partial discharge (PD) inception electric field is higher than the maximum electric field in the Fiber Reinforced Plastic (FRP) core of the composite cross-arm in the fully composite pylon, meanwhile, it is higher than the maximum allowable electric field magnitude in a Fiber Reinforced Plastic (FRP) core of a composite insulator/cross-arm recommended by the industry. It is proved that the partial discharge (PD) performance of these samples meet the requirements in the case of the fully composite pylon.

2.4.3.2 Dielectric Characteristics

From dielectric characteristics test, samples manufactured by hand lay-up moulding method have the lowest relative permittivity, whereas the other two methods lead to a higher and similar relative permittivity.

A model has been proposed to predict the relative permittivity of a fiber glass reinforced polymer composite [65, 66], as shown in Eq. (2.23).

$$\frac{1}{\frac{v_g}{\varepsilon_g}+\frac{v_p}{\varepsilon_p}} \leqslant \varepsilon \leqslant v_g\varepsilon_g + v_p\varepsilon_p \tag{2.23}$$

where v_g and v_p are the fiber volume fraction and polymer volume fraction in a Fiber Reinforced Plastic (FRP) composite respectively, ε_g and ε_p are the relative permittivity of the fiber and the polymer respectively, and ε is the relative permittivity of the Fiber Reinforced Plastic (FRP) material.

It is obvious from Eq. (2.23) that the relative permittivity of a Fiber Reinforced Plastic (FRP) composite depends on the individual relative permittivity and volume fraction of fibers and the resin. It should be noted that those models cannot provide accurate value of relative permittivity of a Fiber Reinforced Plastic (FRP) composite, as it is impossible to measure the accurate volume fractions of included components.

In [65], the relative permittivity of e-glass fiber is measured around 6 while the value for epoxy resin is around 3. We assume that different fibers and resins in Fiber Reinforced Plastic (FRP) samples in question have the same relative permittivity, i.e. $\varepsilon_g = 6$ and $\varepsilon_p = 3$ for simplicity and substitute these values into Eq. (2.23). Then the reason that samples manufactured by hand lay-up method have the lowest relative permittivity may be explained, since they have the lowest fiber volume fraction according to Table 2.2.

The reason why the samples made by the hand lay-up method have a relative high dissipation factor is not clear yet. One guess is that these samples have absorbed moisture since it contains more voids and the moisture increase the conductivity of the bulk material, which leads to a higher power loss. However, the reasons need further investigation.

There is not any officially accepted criterion for dissipation factor or relative permittivity for a solid insulation material. However, a lower power loss is always desired.

Based on all the test results, glass/epoxy manufactured by filament winding is preferable for the application to the Fiber Reinforced Plastic (FRP) tube of the composite cross-arm in the fully composite pylon.

2.5 Electrical-Mechanical Combined Test on Fiber Reinforced Plastic (FRP) Composites

In the operation of the fully composite pylon, the composite cross-arm is exposed to multiple stresses simultaneously, especially mechanical stress and electrical stress.

The mechanical stress is mainly from the dead weight of conductors, wind load, icing load and more critical, from the combination of these loads. The electrical stress is from the operational voltage.

In order to investigate the electrical performance of the Fiber Reinforced Plastic (FRP) core under mechanical stress, an electrical-mechanical combined setup was developed, with which an electrical-mechanical combined test was conducted in the high voltage lab at Department of Electrical Engineering, Technical University of Denmark (DTU). It was a cooperative test performed by project partners Aalborg University (AAU) and Technical University of Denmark (DTU) in the project Power Pylons of the Future (PoPyFu).

In the combined test, a cyclic loading with a frequency of 5 Hz is applied to the specimen until it is broken. It takes long time (from ten thousands of cycles to several million cycles) for each specimen to break. Presently, only 5 specimens have been tested and contrast tests with pure electrical or mechanical stress have not been finished yet. In the present chapter, a brief introduction regarding the combined test concept, the innovative combined setup and initial test results will be given.

2.5.1 Combined Test Circuit and Setup

2.5.1.1 Combined Test Circuit

The schematic of the electrical-mechanical combined test circuit is shown in Fig. 2.22.

The test object is installed in a combined setup and exposed to electrical and mechanical stress simultaneously. The combined setup will be introduced in the following part. The dynamic mechanical load is measured and controlled by Computer1.

With the two stresses acting on the test object, partial discharge (PD) activities from the test object is measured by ICMsystem, a partial discharge (PD) measuring unit manufactured by POWER DIAGNOSTIx. The partial discharge (PD) measuring mechanism is the same with that in Fig. 2.14 and introduced in Sect. 2.4.1.2.

2.5.1.2 Combined Setup

Figure 2.23 indicates the design concept of the combined setup, which introduces electrical and mechanical stress to the specimen simultaneously. The real picture of the combined setup with test object is shown in Fig. 2.24.

The key component of the combined setup is a hydraulic INSTRON universal tensile machine, which has three grips. The tensile machine is able to provide a maximum 100 kN force through the upper and lower grip. The specimen is gripped by the middle and lower grip. There is a strain gauge load cell equipped to increase the

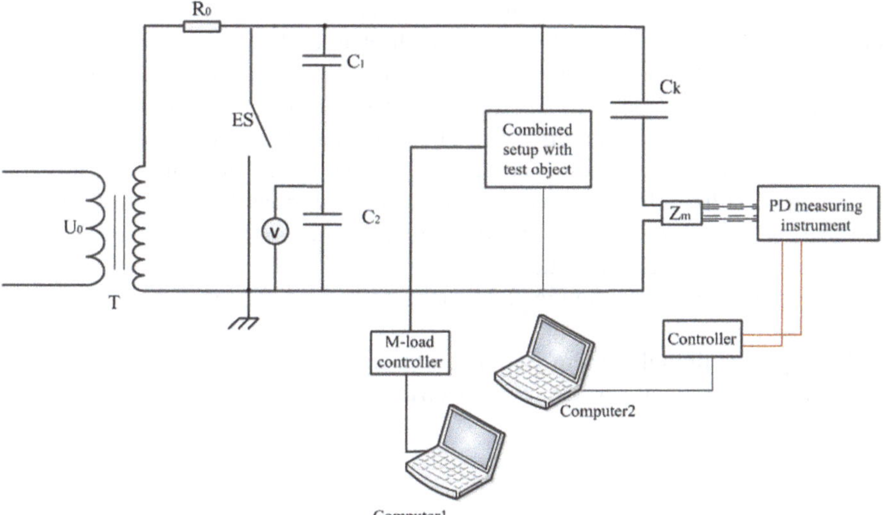

Fig. 2.22 Schematic of the electrical-mechanical combined test circuit

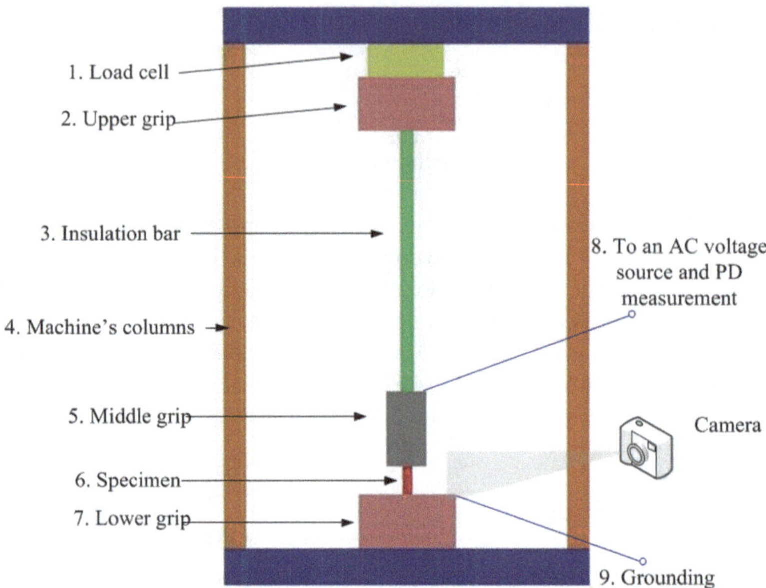

Fig. 2.23 Schematic of the combined setup [67]

fatigue load measuring resolution. The maximum load that the load cell can withstand is 50 kN, i.e. the allowable maximum mechanical load applied to the specimen is 50 kN, instead of 100 kN.

The middle and lower grip also function as HV electrode and ground electrode, respectively. The middle grip is connected to a AC power frequency voltage transformer and to a coupling capacitor for partial discharge (PD) measurement. Between the upper and middle grip (the high voltage (HV) electrode), there is an insulation bar made from glass/epoxy to electrically isolate the upper and middle grip. The lower grip is grounded. There is a risk of surface flashover along the specimen at the instant when the specimen is completely broken by the mechanical load while the voltage is still on. In order to avoid the flashover current flowing into the tensile machine and destroy the load controller, a ground plate is added on the lower grip to collect the possible high current.

The control cable shown in Fig. 2.24 connects the load cell to the mechanical load controller.

An Olympus digital camera is utilized to observe the possible damage evolution from the specimen surface, with electrical and mechanical stress acting on it simultaneously.

1 – Load cell;
2 – Upper grip;
3 – Insulation bar;
4 – Machine's columns;
5 – Middle grip;
6 – Specimen;
7 – Lower grip;
8 – To an AC voltage source amd PD measurement
9 – Grounding plate;
10 – Control cable.

Fig. 2.24 Photo of the combined setup with specimen [67]

2.5.2 Combined Test

2.5.2.1 Test Specimen

The specimens used in the test were E-glass reinforced epoxy (abbreviated as glass/epoxy) and manufactured by SAERTEX through TUCO Marine Aps. The specimen had unidirectional fibers and two laminates configuration with a nominal thickness of 1.8 mm.

Figure 2.25 indicates the configuration and dimension of the specimen. As Fig. 2.25 shows, there are two parts on both ends of the specimen for gripping by middle and lower grips. The part in the middle is for combined testing. Figure 2.25 also indicates the fiber direction in the specimen.

For the time being, 5 specimens have been tested.

2.5.2.2 Electrical and Mechanical Loading

Table 2.3 shows the applied electrical and mechanical loading applied to the specimens in the combined test.

A sinusoidal mechanical loading with a frequency of 5 Hz is applied in the combined test. Different stress magnitude are applied to specimens to obtain effects of different mechanical loads, as Table 2.3 shows.

As introduced in Sect. 2.3.2, the maximum electric field in the Fiber Reinforced Plastic (FRP) core of the composite cross-arm is calculated as $0.6\ \text{kV}_{\text{rms}}/\text{mm}$. Thus, considering the worst case, a voltage $U = 21\ \text{kV}_{\text{rms}}$ should be applied to the specimen. However, it is found out that the controller of the hydraulic tensile machine is sensitive to the electromagnetic interference when surface flashover happens along the specimen. In order to avoid surface flashover that may lead to destroy of the tensile machine controller, a pre-test is performed to obtain the flashover voltage $U_{flashover}$ of all specimens. The test voltage U_0 for each specimen is kept 8% lower than $U_{flashover}$.

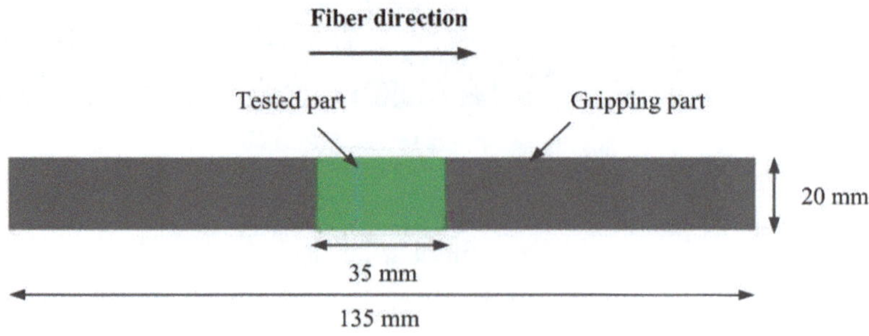

Fig. 2.25 Configuration and dimension of the specimen used in the combined test

Table 2.3 Applied test voltage/electric field and mechanical stress magnitude

Number	$U_{flashover}$ [kV_{rms}]	U_0 [kV_{rms}]	d [mm]	E_0 [kV_{rms}/mm]	Stress magnitude [MPa]
1#	22.50	21.00	Not measured (<35.00)	(<0.60)	210
2#	19.00	17.50	32.10	0.55	200
3#	24.60	22.60	33.03	0.68	180
4#	25.00	23.00	35.50	0.65	150
5#	19.40	17.80	31.00	0.57	150

Although the designed length of the tested part in a specimen is 35 mm, the distance *d* between the middle and lower grip varies in practice. The practical applied electric field E in the specimen is calculated by $E_0 = U_0/d$. Table 2.3 shows that electric fields applied to all specimens are not identical and there is a deviation with a maximum value of 20% approximately. The applied voltage is a standard sinusoidal one with power frequency.

2.5.2.3 Test Procedure

The dynamic mechanical loading and the voltage are applied to the specimen simultaneously until fatigue failure happens in the end of the test. A fatigue failure indicates

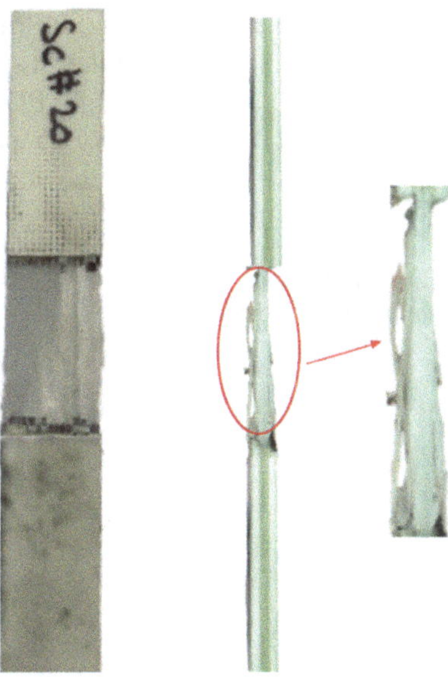

Fig. 2.26 A broken specimen from front view (left) and side view (right) with zoom-in details

a localized permanent structure change of a specimen and it can no longer withstand the tensile force. Figure 2.26 gives an example of a failed specimen where fibers in the specimen are partially broken. If the specimen does not fail after 1.5 million cycles, the test stops and the corresponding specimen is recorded as runout.

During the whole process, partial discharge (PD) activities are recorded and the surface damage evolution on the specimen, such as forming of cracks and fractures is observed by the digital camera.

2.5.3 *Test Results*

Table 2.4 shows number of mechanical loading cycles before a specimen was broken. With decrease of mechanical stress, cycles to failure increases. The specimen 5# did not experience failure after 1.5 million cycles and it was recored as 'runout'. With the same stress, specimen 4# was broken whereas 5# was not. One reason is that the low stress is close to the fatigue limit, with which the material will never break. With this low stress, the random variation in the material quality leads to a difference in the failure cycles.

Through observations, partial discharge (PD) activities from all specimens under different mechanical stresses have similar development trending.

Figure 2.27 shows typical temporal development of partial discharge (PD) magnitude over time. In the figure, the abscissa represents time in second and the ordinate represents partial discharge (PD) magnitude in pC. The red curve in the figure indicates average partial discharge (PD) magnitude per second, while the blue curve indicates the applied voltage. Each sub-figure indicates the partial discharge (PD) development over a period of 8 min. The first sub-figure (t = 8 min) indicates the first 8 min of the test. The last sub-figure (t = 376 min) indicates the last 8 min of the test, when the specimen broke. It should be noted that there was a measuring limit of 576 pC for partial discharge (PD) magnitude during the test.

The typical development of the partial discharge (PD) magnitude during the combined test may be divided into 3 stages roughly.

Table 2.4 Number of cycles a specimen went through before failure

Specimen	Stress magnitude [MPa]	Cycles to failure
1#	210	60,000
2#	200	110,000
3#	180	300,000
4#	150	500,000
5#	150	1,500,000 (runout)

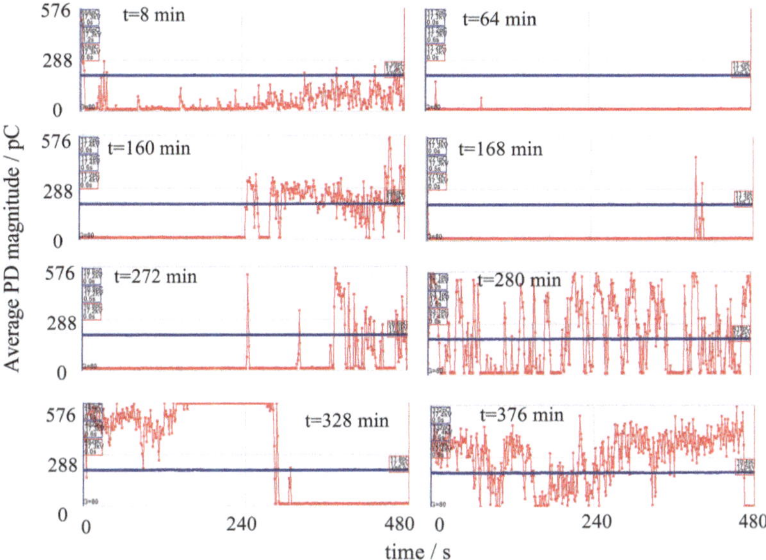

Fig. 2.27 Typical temporal development of partial discharge (PD) magnitude for specimen 2#: the blue curve indicates applied voltage and red curve indicates average partial discharge (PD) magnitude per second

In the first stage, i.e. as soon as the mechanical and electrical stress were applied, partial discharge (PD) was triggered from the specimen and the partial discharge (PD) level was not high. After thousands of cycles, for example after 64 min (19200 cycles) in Fig. 2.27, the partial discharge (PD) activity was extinct. One possible source for the first stage partial discharge (PD) signals may be the impurities, such as grease or dust, remained on the specimen surface. Those impurities may be removed by the electrical discharge and then the partial discharge (PD) were extinct. In the future research, a through cleaning of the specimen should be done before the test.

In the second stage, bigger partial discharge (PD) signals started to appear and disappear abruptly, as the third (t = 160 min) to fifth (t = 272 min) sub-figure in Fig. 2.27 show. In this stage, the interval between partial discharge (PD) extinction and the following initiation seems to be random.

As it was close to the specimen failure, the partial discharge (PD) magnitude increased obviously and the time interval between partial discharge (PD) extinction and the following initiation decreased. This stage can be regarded as the final stage of the partial discharge (PD) development in the combined test, as the sixth (t = 280 min) to eighth (t = 376 min) sub-figure in Fig. 2.27 show.

For specimen 2#, the average partial discharge (PD) magnitude was lower than 200 pC in the first stage of partial discharge (PD) development. In the second stage, this value increased to 300–400 pC approximately. When it was close to specimen failure, the continuous partial discharge (PD) signal had a magnitude larger than 400 pC, and sometimes higher than 600 pC considering the measuring limit. For

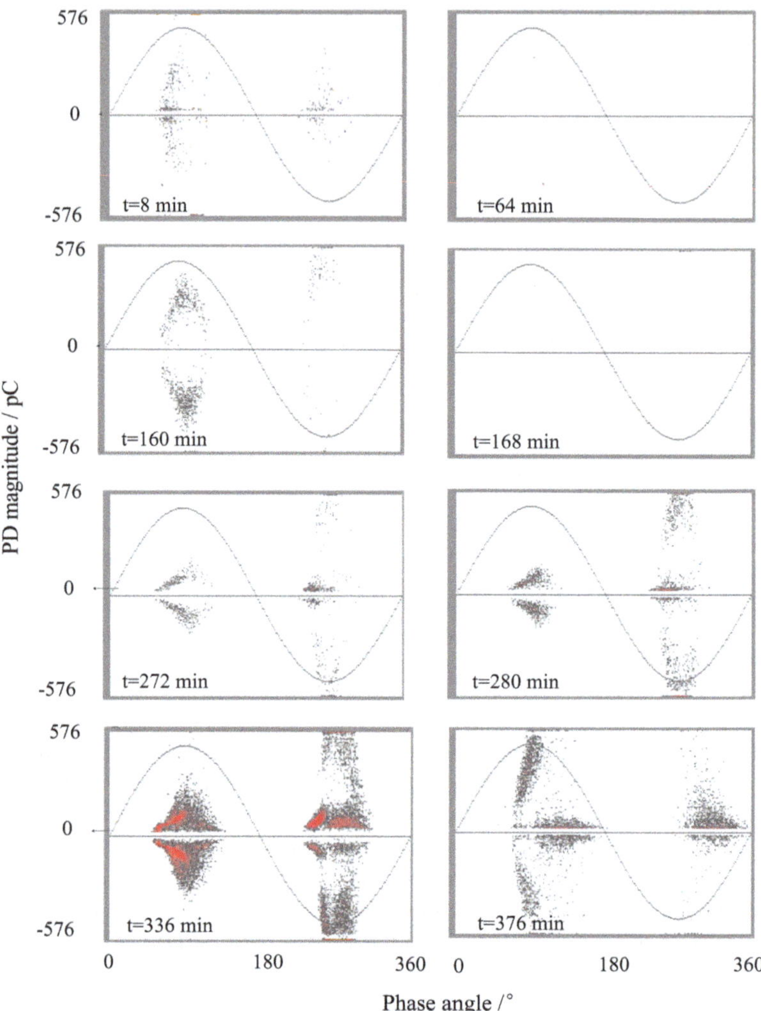

Fig. 2.28 Typical temporal development of phase resolved partial discharge (PRPD) for specimen 2#

other specimens, the partial discharge (PD) signal experienced the same trend but with different partial discharge (PD) magnitudes. The increase of partial discharge (PD) magnitude and repetitive rate may be utilized as an indicator for the specimen failure. However, until now only a small number of specimens have been tested, more data are needed for significant conclusions.

Figure 2.28 shows typical temporal development of phase resolved partial discharge (PRPD) patterns over time.

From the phase resolved partial discharge (PRPD) patterns shown in Fig. 2.28, the measured partial discharge (PD) is more like surface discharge, where substantial asymmetry is observed with positive and negative voltage. Thus, it is rational to have the hypothesis that the development of partial discharge (PD) activities during the combined test is affected by the surface condition of the specimen. With mechanical dynamic loading, surface damage, such as small cracks may be formed on the specimen surface and they may lead to changing of partial discharge (PD) magnitude, partial discharge (PD) patterns and repetitive rate. However, photos taken by the digital camera during the test on all specimens did not show any sign of surface degradation before the specimen was totally broken. It is possible that the micro crack on the specimen is too small to be recored by a normal digital camera.

2.5.4 Discussion

From the partial discharge (PD) development over time during the electrical-mechanical test, it is inferred that mechanical loading have effects on partial discharge (PD) activities from Fiber Reinforced Plastic (FRP) materials, since all specimens experienced a partial discharge (PD) increase when they were close to mechanical failure. An effective way to prove the effects of mechanical loading on partial discharge (PD) activities is to conduct a contrast test— applying electrical stress only to the specimen and comparing partial discharge (PD) activities in pure electrical tests with that in the combined test. Due to time and facility limitation, the pure electrical test on these specimens have not been finished until the book is complete.

The mechanism regarding effects of mechanical load on electrical performance of Fiber Reinforced Plastic (FRP) materials are not clear for the time being. One hypothesis is that the dynamic loading causes micro cracks on the specimen surface and these cracks leads to changing of surface discharge.

Due to time limitation, only 5 specimens have been tested for the time being. With these limited test results, only limited conclusions can be drawn. More specimens need to be tested to obtain significant conclusions.

Moreover, for the time being, the electric field and tensile force in the combined test are both along the fiber direction, which does not represent all situations in service. In the future, directions of electric field and tensile force perpendicular to fibers should also be considered.

Although the combined test results presented here are not yet sufficient to conclusions of great significance, the electrical-mechanical combined test is an useful method in the validation of Fiber Reinforced Plastic (FRP) core of the composite cross-arm in the fully composite pylon. With the combined test setup after necessary development, the following outputs are expected in the future

- Effects and mechanism of mechanical loading on partial discharge (PD) activities from the Fiber Reinforced Plastic (FRP) core, considering different electric field and mechanical force directions with respect to the fiber direction;

- Effects and mechanism of electric field on the lifetime of Fiber Reinforced Plastic (FRP) materials, considering different electric field and mechanical force directions with respect to the fiber direction;
- Mechanism of aging and degradation of Fiber Reinforced Plastic (FRP) materials caused by electrical-mechanical combined stress;
- Development of means for predicting the mechanical failure of Fiber Reinforced Plastic (FRP) materials by partial discharge (PD) measurement.

What's more, with the combined test concept, a combined test setup is designed in the project Power Pylons of the Future (PoPyFu) to test the full-scale composite cross-arm with simultaneous electrical-mechanical stresses. The setup will not be discussed here.

2.6 Summary

In this chapter, two candidate Fiber Reinforced Plastic (FRP) materials—glass/epoxy and glass/vinyl ester samples, manufactured by three different methods—hand lay-up, vacuum assisted resin transfer moulding and filament winding- have been tested, which will be used as the Fiber Reinforced Plastic (FRP) core material of the composite cross-arm in the fully composite pylon. The internal partial discharge (PD) performance, dissipation factor and relative permittivity of these Fiber Reinforced Plastic (FRP) materials are considered. Test results proved that the glass/epoxy sample, made by filament winding, have the highest partial discharge (PD) inception electric field, lowest dissipation factor and relative high value of relative permittivity.

Based on the test results, the glass/epoxy manufactured by filament winding should be regarded as the suitable candidate material for Fiber Reinforced Plastic (FRP) core of the composite cross-arm in the fully composite pylon.

Additionally, an innovative electrical-mechanical combined test setup is introduced in this chapter. With the combined setup, the Fiber Reinforced Plastic (FRP) sample can be exposed to electrical and mechanical stress simultaneously. Partial discharge (PD) activities and real-time surface damage evaluation from the Fiber Reinforced Plastic (FRP) sample can be monitored during the combined test. In this chapter, initial results indicating mechanical effects on partial discharge (PD) activities of Fiber Reinforced Plastic (FRP) materials are introduced. More details and full results regarding electrical-mechanical combined tests on Fiber Reinforced Plastic (FRP) materials will be published in [67].

The combined test is an useful method to investigate effects of mechanical loading on electrical behaviors of Fiber Reinforced Plastic (FRP) materials in the fully composite pylon, which is always stressed by electrical and mechanical effects in service. In return, effects of electric field on the mechanical performance of the Fiber Reinforced Plastic (FRP) material can also be studied. More tests will be performed for this topic in the future.

References

1. L.C. Hollaway, A review of the present and future utilisation of FRP composites in the civil infrastructure with reference to their important in-service properties. Constr. Build. Mater. **24**(12), 2419–2445 (2010)
2. T.P. Sathishkumar, S. Satheeshkumar, J. Naveen, Glass fiber-reinforced polymer composites—a review. J. Reinf. Plast. Compos. **33**(13), 1258–1275 (2014)
3. M.N. Gururaja, A.N. Rao, A review on recent applications and future prospectus of hybrid composites. Int. J. Soft Comput. Eng. (IJSCE) **1**(6), 352–355 (2012)
4. A. Avci, H. Arikan, A. Akdemir, Fracture behavior of glass fiber reinforced polymer composite. Cem. Concr. Res. **34**(3), 429–434 (2004)
5. A.M. Visco, L. Calabrese, P. Cianciafara, Modification of polyester resin based composites induced by seawater absorption. Compos. Part A Appl. Sci. Manuf. **39**(5), 805–814 (2008)
6. P.K. Mallick, *Fiber-reinforced composites: materials, manufacturing, and design* (CRC Press, 2007)
7. T.K. Sørensen, Composite based EHV AC overhead transmission lines, Ph.D. dissertation, Technical University of Denmark (2010)
8. N. Gupta, B.S. Brar, E. Woldesenbet, Effect of filler addition on the compressive and impact properties of glass fibre reinforced epoxy. Bull. Mater. Sci. **24**(2), 219–223 (2001)
9. B. Suresha, G. Chandramohan, J.N. Prakash, V. Balusamy, The role of fillers on friction and slide wear characteristics in glass-epoxy composite systems. J. Miner. Mater. Charact. Eng. **5**(01), 87 (2006)
10. L. Zhang, Q. Sun, X. Zhao, H. Wang, Material selection for transmission tower made of fiber reinforced plastics. Electr. Power Constr. **32**(2), 1–5 (2011). (in Chinese)
11. M.F.G. Companies, Technical design guide for FRP composite products and parts, http://www.moldedfiberglass.com/resources
12. J. Toth, Transmission line structures with fiber reinforced polymer (FRP) composites, SCB2 Transmission Lines Tutorial Session in CIGRE SCC3 meeting in Seoul joined with SCB2-Electric Environment Technology Colloquium, Jun 2017
13. H.Y. Yeh, S.C. Yang, Building of a composite transmission tower. J. Reinf. Plast. Compos. **16**(5), 414–424 (1997)
14. S. Deng, H. Li, Q. Wei, G. Hu, and K. Liu, Material, electrical and mechanical characteristics tests of composite pole for 110 kV overhead transmission lines. South. Power Syst. Technol. **5**(3), 36–40 (2011). (in Chinese)
15. M. Selvaraj, S.M. Kulkarni, R.R. Babu, Behavioral analysis of built-up transmission line tower from FRP pultruded sections. Int. J. Emerg. Technol. Adv. Eng. **2**(9), 39–47 (2012)
16. J. Massy, *Thermoplastic and Thermosetting Polymers in a Little Book About Big Chemistry* (Springer, Cham, 2017)
17. G. Yilmaz, O. Kalenderli, Dielectric properties of aged polyester films, in *Conference on Electrical Insulation and Dielectric Phenomena: Annual Report*, vol. 2, Oct 1997, pp. 444–446
18. H.S. McNaughton, Comparative performance of woven glass and polyester film backed micaceous tapes for the insulation of high voltage AC stator windings of rotating electrical machines, in *Conference on Electrical/Electronics Insulation*, Chicago, USA, Oct 1993, pp. 545–551
19. S. Ibrahim, D. Polyzois, S.K. Hassan, Development of glass fiber reinforced plastic poles for transmission and distribution lines. Can. J. Civ. Eng. **27**(5), 850–858 (2000)
20. K.N. Shivakumar, G. Swaminathan, M. Sharpe, Carbon/vinyl ester composites for enhanced performance in marine applications. J. Reinf. Plast. Compos. **25**(10), 1101–1116 (2006), https://doi.org/10.1177/0731684406065194
21. E. Zghayar, K. Mackie, J. Xia, Fiber reinforced polymer (FRP) composites in retrofitting of concrete structures: polyurethane systems versus epoxy systems. Int. Concr. Abstr. Portal **298**, 1–22 (2014)
22. C. Wang, P. Yue, Z. Liu, Z. Xia, Electrical properties study of epoxy resin composite material applied in transmission tower. Insul. Mater. **47**(2), 36–40 (2014)

23. G. McDaniel, C. Knight, Design training expo-fiber reinforced polymer (FRP) composites (2014), https://www.slideshare.net/AravindGanesh1/fibre-reinforced-polymerfrp
24. J.F. Hall, History and bibliography of polymeric insulators for outdoor applications. IEEE Trans. Power Deliv. **8**(1), 376–385 (1993)
25. P. Dey, B.J. Drinkwater, S.H.R. Proud, Developments in insulation for high voltage overhead transmission systems, in *9th Conference on Electrical Insulation*, Boston, USA, Sep 1969, pp. 38–43
26. C. W. 22-03, Worldwide service experience with HV composite insulators. Electra **130**, 68–77 (1990)
27. R. Hackam, Outdoor HV composite polymeric insulators. IEEE Trans. Dielectr. Electr. Insul. **6**(5), 557–585 (1999)
28. X. Liang, S. Wang, J. Fan, Z. Guan, Development of composite insulators in China. IEEE Trans. Dielectr. Electr. Insul. **6**(5), 586–594 (1999)
29. J. Mackevich, M. Shah, Polymer outdoor insulating materials. Part I: comparison of porcelain and polymer electrical insulation. IEEE Electr. Insul. Mag. **13**(3), 5–12 (1997)
30. T. Sorqvist, A.E. Vlastos, Performance and ageing of polymeric insulators. IEEE Trans. Power Deliv. **12**(4), 1657–1665 (1997)
31. CIGRE, *CIGRE Green Book on Overhead Lines* (Paris, 2014)
32. EPRI, Epri database, presented at the WG-meeting of CIGRE B2. 21 in Winterback (2008)
33. J.T. Burnham, T. Baker, A. Bernstorf, C. de Tourreil, J. George, R. Gorur, R. Hartings, B. Hill, A. Jagtiani, T. McQuarrie, D. Mitchell, D. Ruff, H. Schneider, D. Shaffner, J. Yu, J. Varner, Task force report: Brittle fracture in nonceramic insulators. IEEE Power Eng. Rev. **22**(5), 71 (2002)
34. H. Okamoto, Y. Ikeda, Arc resistance and application of FRP to arms in overhead power-line towers. IEEE Trans. Power Appar. Syst. **9**, 1098–1102 (1967)
35. J.N. Hoffmann, R.W. Wiedmer, M.J. Bubniak, I.S. Moreira, Urban overhead transmission lines of compact design for 69, 138 and 230 kV, in *Cigré Session*, Paris, France, Aug 2010
36. D.I. Lee, K.Y. Shin, S.Y. Lee, W.K. Lee, EMF mitigation characteristics of 154 kV compact transmission tower using insulation arms, in *CIGRE Session*, Paris, France, Aug 2010
37. F.S.K. Papailiou, W. Van, J.F.P.J. Kolemeijer, Further developments of compact lines for 420 kV with silicone insulators and their advantages for applications in emergency restoration systems, Paris, France, Aug 2004
38. TenneT, Wintrack ii high-voltage pylons, https://www.tennet.eu/news/detail/tennet-awards-design-and-construction-of-innovative-wintrack-ii-high-voltage-pylons-to-heijmans-euro/
39. C. Zachariades, I. Cotton, S.M. Rowland, V. Peesapati, P.R. Green, D. Chambers, M. Queen, A coastal trial facility for high voltage composite cross-arms, in *IEEE International Symposium on Electrical Insulation*, San Juan, US, Jun 2012, pp. 78–82
40. C. Zachariades, S.M. Rowland, I. Cotton, V. Peesapati, D. Chambers, Development of electric-field stress control devices for a 132 kV insulating cross-arm using finite-element analysis. IEEE Trans. Power Deliv. **31**(5), 2105–2113 (2016)
41. X. Yang, N. Li, Z. Peng, J. Liao, Q. Wang, Potential distribution computation and structure optimization for composite cross-arms in 750 kV AC transmission line. IEEE Trans. Dielectr. Electr. Insul. **21**(4), 1660–1669 (2014)
42. A.C. Baker, R.A. Bernstorf, E. del Bello, R.J. Hill, B. King, A.J. Phillips, D.G. Powell, D. Shaffner, G.A. Stewart, T. Grisham, Guide for braced insulator assemblies for overhead transmission lines 60 kV and greater. IEEE Trans. Power Deliv. **23**(2), 785–791 (2008)
43. G.A. Escher, *Design and Testing of an Optimum Fibreglass Reinforced Pole and market study on the Demand for Utility Poles in Canada* (Canadian Electrical Association, 1982)
44. M. Sarmento, B. Lacoursiere, A state of the art overview: composite utility poles for distribution and transmission applications, in *Transmission and Distribution Conference and Exposition: Latin America*, Caracas, Venezuela, Aug 2006, pp. 1–4
45. H.M. Li, S.C. Deng, Q.H. Wei, Y.N. Wu, Research on composite material towers used in 110 kV overhead transmission lines, in *International Conference on High Voltage Engineering and Application*, New Orleans, USA, Oct 2010, pp. 572–575

46. R. T. Inc. Composite utility pole, http://www.rspoles.com/resources/photos/transmission
47. I. S. 60270, *High voltage test techniques—partial discharge measurements*, Std. (2000)
48. A. Kutil, K. Frohlich, Partial discharge phenomena in composite insulation materials, in *Conference on Electrical Insulation and Dielectric Phenomena*, Virginia Beach, USA, Oct 1995, pp. 343–346
49. A. Kutil, Qualification of fiber-reinforced insulating materials based on PD analysis. IEEE Trans. Dielectr. Electr. Insul. **5**(4), 603–611 (1998)
50. A. Kutil, K. Frohlich, A multistress test procedure for qualification of composite insulation materials in GIS, in *IEEE International Symposium on Electrical Insulation*, Pittsburgh, USA, Jun 1994, pp. 58–61
51. Z. Jia, Y. Hao, H. Xie, The degradation assessment of epoxy/mica insulation under multi-stresses aging. IEEE Trans. Dielectr. Electr. Insul. **13**(2), 415–422 (2006)
52. H.P. Burgener, K. Frohlich, "Impact of mechanical stress on the dielectric behaviour of fibre-reinforced insulation materials, in *11th International Symposium on High Voltage Engineering*, vol. 4, London, UK, Aug 1999, pp. 356–359 vol.4
53. I. S. 60250, *Recommended methods for the determination of the permittivity and dielectric dissipation factor of electrical insulating materials at power, audio and radio frequencies including metre wavelengths*, Std. (1969)
54. E. Kuffel, W.S. Zaengl, J. Kuffel, *High voltage engineering—fundamentals*, 2nd edn. (Butterworth-Heinemann, 2000). Chapter 4: Electrostatic fields and field stress control, p. 203
55. I. S. 60296, *Fluids for electrotechnical applications-unused mineral insulating oils for transformers and switchgear*, Std. (2012)
56. E. Kuffel, W.S. Zaengl, J. Kuffel, *High voltage engineering—fundamentals*, 2nd edn. (Butterworth-Heinemann, 2000). Chapter 7: Non-destructive insulation test techniques, p. 419
57. T. Jahangiri, Q. Wang, C.L. Bak, F.F. Silva, Electric stress computations for designing a novel unibody composite cross-arm using finite element method. IEEE Trans. Dielectr. Electr. Insul. **24**(6), 3567–3577 (2017)
58. ABB, Hollow composite insulators (72–1200 kV), https://new.abb.com/products/transformers/transformer-components/composite-insulators/hollow-composite-insulators-(72---1-200-kv)
59. D. Gay, S.V. Hoa, S.W. Tsai, *Composite materials: design and applications* (CRC Press, 2002)
60. Y.K. Hamidi, S. Dharmavaram, L. Aktas, M.C. Altan, Effect of fiber content on void morphology in resin transfer molded e-glass/epoxy composites. J. Eng. Mater. Technol. **131**(2), 021014 (2009)
61. L.D. Landro, A. Montalto, P. Bettini, S. Guerra, Detection of voids in carbon epoxy laminates and their influence on mechanical properties. Polym. Polym. Compos. **25**(5), 371 (2017)
62. C.J. DeValve, Investigations on void formation in composite molding processes and structural damping in fiber-reinforced composites with nanoscale reinforcements, Ph.D. dissertation, Virginia Tech (2013)
63. S. He, Y. Wen, W. Yu, H. Liu, Study on voids of epoxy matrix composites sandwich structure parts, in *IOP Conference Series: Materials Science and Engineering*, vol. 182, no. 1 (IOP Publishing, 2017), p. 012031
64. S. Hernández, F. Sket, J.M. Molina, C. González, Effect of curing cycle on void distribution and interlaminar shear strength in polymer-matrix composites. Compos. Sci. Technol. **71**(10), 1331–1341 (2011)
65. V.I. Sokolov, S.I. Shalgunov, I.G. Gurtovnik, L.G. Mikheeva, Dielectric characteristics of glass fibre reinforced plastics and their components. Int. Polym. Sci. Technol. **32**(7), T62 (2005)
66. M. Dinulović, B. Rašuo, Dielectric properties modeling of composite materials. FME Trans. **37**(3), 117–122 (2009)
67. M. Manouchehr, Q. Wang, C. Berggreen, J. Holboll, C.L. Bak, Combined mechanical-electrical fatigue of UD glass fiber composites. (in preparation)

Chapter 3
Air Clearances of Fully Composite Pylon

3.1 Introduction

One of the most important requirements in the electrical design of power pylons is the determination of air clearances between grounded parts and live parts. The air clearances can be categorized into internal and external clearances. Internal clearances are the distances between two phases' conductors, as well as between phase conductors and grounded tower structure/shield wires. External clearances cover the distances between phase conductors and other external adjacent elements such as trees, railways, vehicles, building, and people.

The air clearances on any high voltage overhead line pylon must be enough to withstand against power frequency, switching and lightning overvoltages. To determine such air clearances, different regulations and guidelines have been adopted by different countries. The National Electrical Safety Code (NESC) [1] applies in USA whereas IEC standard 60071 [2, 3] is utilized in Europe. BS-EN 50341 [4, 5] together with IEC 60071 are used in UK [6]. However, insulation coordination studies should still be done to determine required air clearances. The insulation coordination procedure and its application guidelines are presented in IEC 60071-1 [2] and IEC 60071-2 [3], respectively. The insulation coordination guidelines by IEEE standard are also described in IEEE Std. 1313.1 [7], 1313.2 [8] and 1427 [9]. The methodologies of insulation coordination by IEEE Standard 1313.1 and IEC 60071-1 are compared in [10] and it is concluded that both standards provide a common line of action in the insulation coordination study. For the purpose of overhead line insulation, the insulation coordination procedures described in IEC Std. 60071-1, 2 and BS-EN 50341-1 contain more details and have been used in [11] to determine minimum required air clearances.

BS EN 50341-1 covers the general specifications of overhead lines exceeding 45 kV. In Annex E of this standard, the formulas proposed by the Central Research Institute of Electric Power Industry (CRIEPI) are presented to compute the required

T. Jahangiri et al., *Electrical Design of a 400 kV Composite Tower*, Lecture Notes in Electrical Engineering 557,
https://doi.org/10.1007/978-3-030-17843-7_3

minimum phase-to-earth and phase-to-phase air clearances on overhead lines [11, 12]. Moreover, the calculation of air clearances based on swing angles of insulators is reported in CIGRÉ TB 348 [4] for various loading conditions such as temperature, wind, and ice at the tower top and mid-span. In this chapter, the insulation coordination study to determine minimum required air clearances is presented and the obtained results for the basic design of a fully composite pylon are discussed.

3.2 Insulation Coordination

Definition of insulation coordination based on IEC Std. 60071-1 is as follows:

"Selection of the dielectric strength of equipment (its rated or its standard insulation level) in relation to the operating voltages and overvoltages which can appear on the system for which the equipment is intended and taking into account the service environment and the characteristics of the available preventing and protective devices" [2].

The main purpose of insulation coordination is to make sure that any insulation failure in the system is in the form of self-restoring and the failure probability falls within the acceptable range [6]. Based on insulation coordination study, the electrical strength of fully composite pylon can be properly selected and designed to provide acceptable and reliable performance for the pylon. This study allows decreasing insulation failures together with reducing the cost of pylon design by avoiding oversizing of the pylon. Insulation coordination has two different approaches: statistical approach and deterministic approach.

Statistical approach: In this approach, a comprehensive knowledge of statistical distribution of insulation strengths and overvoltages are required to determine the risk of insulation failure. This approach cannot be used to evaluate the electrical strength of fully composite pylon, because fully composite pylons do not exist in the real world and there is no experience regarding their actual performance.

Deterministic approach: This approach is referred in IEC Std. 60071. In this approach, it is assumed that all of the maximum possible overvoltages and the insulation strength of equipment of a system are at a certain level. This level can be found by either system simulations or design experiences and test measurements. The deterministic approach has been used in [11] and [13] for insulation coordination study of fully composite pylon. In [11], the transient overvoltages of a system composed of fully composite pylons are assumed to be at specific levels and provide the same dielectric effect on the insulation, as withstand voltages of a given standard switching impulse withstand voltage (SIWV) and lightning impulse withstand voltage (LIWV) defined in IEC Std. 60071-1, 2.

3.2.1 Overvoltages

The transient overvoltages, which can occur in the system due to different origins, can be replaced by representative overvoltages. The representative overvoltages have standard shapes and durations as specified in Table 3.1. The slow front (sf) overvoltages mainly represent the overvoltages due to switching operations and the fast front (ff) overvoltages are related to the overvoltages caused by lightning strikes. Power frequency (pf) overvoltages cover the temporary overvoltages (TOV) due to faults or switching operations (such as resonance conditions and load rejection) in the system. The amplitudes of these overvoltages depend on the type of overvoltage and are important in the choice of the required withstand voltage level for the equipment insulation and air gap insulation. Fast front and slow front overvoltages are usually determinant in the definition of required air clearances for the overhead lines. In IEC Std. 60071-1, two categories have been considered for the highest system voltage of equipment (U_s) in which the determinant impulse voltages for the selection of insulation level are mentioned [14]:

Range I: for U_s above 1 to 245 kV included; insulation level determines according to lightning impulses (fast front), U_{ff}

Range II: for U_s above 245 kV; insulation level specifies based on switching impulses (slow front), U_{sf}

In IEC Std. 60071-2, the guidelines to determine the rated withstand voltages for the two voltage ranges of I and II are presented to justify the association of these rated withstand voltages with the standardized highest system voltages for equipment [3].

3.2.2 Insulation Strength Characteristics

Air is the major insulation in overhead transmission lines. The main dielectric characteristic of air is its disruptive discharge that may occur due to applying voltage stresses. Air insulation breakdown mainly depends on gap configuration, waveform and polarity of applied voltage stress. Atmospheric condition has also influence on the breakdown strength of air. Air insulation breakdown process has a statistical feature with self-restoring capability.

Typically, self-restoring insulation is characterized by a statistical withstand voltage which is corresponding to withstand probability of 90%[1] [3]. Conversely, non-self-restoring insulation such as glass insulator does not have statistical nature and its actual withstand voltage is corresponding to withstand probability of 100%.[2]

[1] Statistical withstand voltage: A self-restoring insulation such as air has 90% withstand probability when the 90% of applied impulse voltages withstand against disruptive discharges and only 10% of the impulse voltages lead to self-restoring insulation failures [3].

[2] Statistical withstand voltage: For a non-self-restoring insulation, the number of disruptive discharges should be zero for a specific conventional withstand voltage (withstand probability of 100%).

Table 3.1 Shapes and durations of voltages and overvoltages and corresponding standard shapes and durations [2, 15]

Class	Low frequency		Transient	
	Continuous	Temporary	Slow-front	Fast-front
Voltage or over-voltage shapes	1/*f* T_t	1/*f* T_t	T_p T_2	T_1 T_2
Range of voltage or over-voltage shapes	*f* = 50 Hz or 60 Hz $T_t \geq 3\ 600$ s	10 Hz < *f* < 500 Hz $0{,}03\ \text{s} \leq T_t \leq 3\ 600$ s	$20\ \mu s < T_p \leq 5\ 000\ \mu s$ $T_2 \leq 20$ ms	$0{,}1\ \mu s < T_1 \leq 20\ \mu s$ $T_2 \leq 300\ \mu s$
Standard voltage shapes	1/*f* T_t *f* = 50 Hz or 60 Hz T_t 1)	1/*f* T_t 48 Hz ≤ *f* ≤ 62 Hz T_t = 60 s	T_p T_2 $T_p = 250\ \mu s$ $T_2 = 2\ 500\ \mu s$	T_1 T_2 $T_1 = 1{,}2\ \mu s$ $T_2 = 50\ \mu s$
Standard withstand test	1)	Short-duration power frequency test	Switching impulse test	Lightning mpulse test

1) To be specified by the relevant apparatus committees.

For a self-restoring insulation, when a specific number of impulse voltages with a given waveform are applied between the electrodes of an air gap, the insulation breakdown may only occur by some of them and the insulation withstands against the rest of impulse voltages.

For this reason, the insulation withstand can be described by a probability function. Two probability functions are presented in IEC Std. 60071-2 which are Gaussian and Weibull functions. Gaussian distribution and modified Weibull distribution are

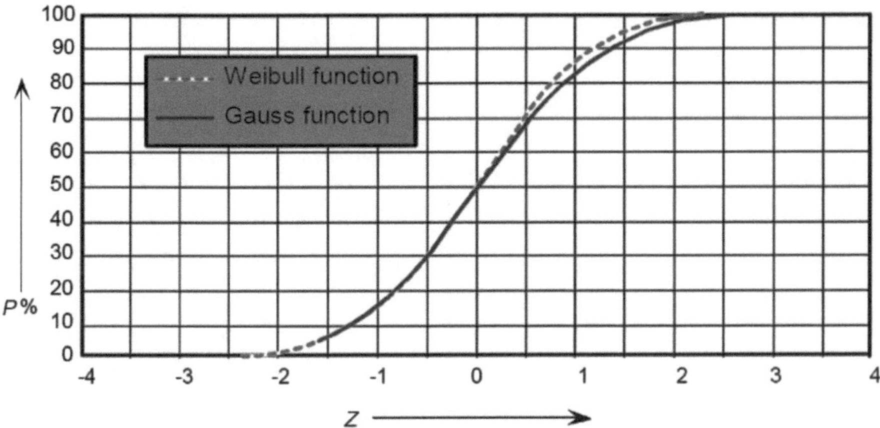

Fig. 3.1 Distributive discharge (flashover) probability of self-restoring insulation based on a linear scale [3]

shown in Fig. 3.1 for a linear scale. In Fig. 3.1, Z is the conventional deviation and indicates the scattering of flashover voltages. Conventional deviation of Z implies to the ability of self-restoring insulation to withstand against impulse voltage stresses in statistical terms with a deviation factor. The recommended values for Z in insulation coordination study are as follows [3, 14]:

For lightning impulse voltages (variation coefficient of 3%): $Z = 0.03\,U_{50}$ (kV)
For switching impulse voltages (variation coefficient of 6%): $Z = 0.06\,U_{50}$ (kV)

where, U_{50} is corresponding to the voltage with which the probability of (disruptive discharge) flashover in insulation is 50%.

3.2.3 *Failure Risk of Insulation*

An acceptable risk of failure due to transient overvoltages should be determined for the electrical strength of insulation, which can be found based on technical and economic aspects or service experiences. The risk of failure indicates the insulation failure probability. The insulation failure rate is the expected average frequency of insulation failures due to overvoltage stresses and can be calculated by multiplying the risk of failure to the total number of overvoltage events. Acceptable failure rates due to overvoltages for apparatus are in the range between 0.001/year and 0.004/year whereas acceptable failure rates due to lightning for overhead lines are between 0.1/100 and 20/100 km/year (the highest number is for distribution lines) [3]. Acceptable failure rates due to switching overvoltages vary in the range between 0.01 and 0.001 per operation. When the cumulative distribution of overvoltages together

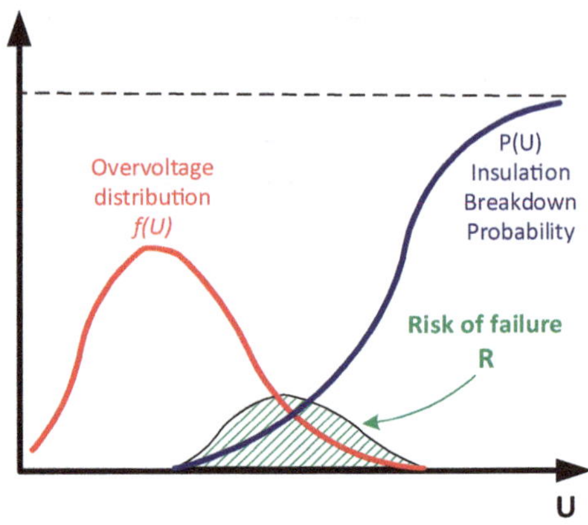

Fig. 3.2 Evaluation of risk of failure [3]

with corresponding the breakdown probability distribution of insulation are known, the failure risk of insulation (shown in Fig. 3.2) can be calculated by [3]:

$$R = \int f(U) \times P(U)dU \quad (3.1)$$

where,

$f(U)$: Probability density of overvoltages (overvoltage occurrence distribution)
$P(U)$: Insulation breakdown probability under an impulse voltage of U

Practically, the risk of failure for a self-restoring insulation is inevitable because it is not feasible to select an insulation at which two curves of $f(U)$ and $P(U)$ do not overlap. Moreover, insulation failure occurs if overvoltage becomes higher than withstand voltage.

In the case that actual distribution of overvoltages and insulation withstand strengths (or breakdown probability) are not known, the risk of failure can be estimated using simplified statistical approach [16]. In this approach, it is assumed that the distribution of overvoltages and insulation withstand strengths are defined using a point on each of distribution curves shown in Fig. 3.3. As shown in Fig. 3.3a, the distribution of insulation withstand strength is identified by a statistical withstand voltage. This voltage is a lightning or switching impulse voltage at which the insulation exhibits a 90% withstand probability.

Statistical withstand voltage[3] implies to the probability of insulation breakdown of 10%. Figure 3.3b also shows that overvoltage distribution is identified by a statistical overvoltage[4] that has a 2% probability of being exceeded.

[3] Applied impulse voltage that 90% of time does not lead to insulation breakdown.

[4] This overvoltage represents probability of occurrence of overvoltages of such amplitude that only 2% has a chance to cause insulation breakdown.

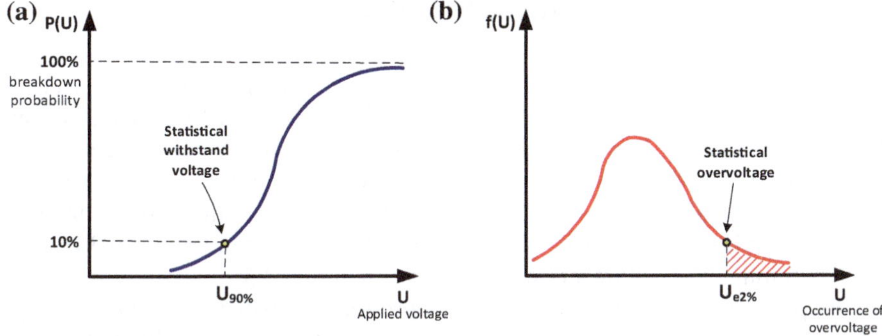

Fig. 3.3 **a** Cumulative distribution function of insulation withstand voltage (strength), **b** Probability density function of overvoltages (stress)

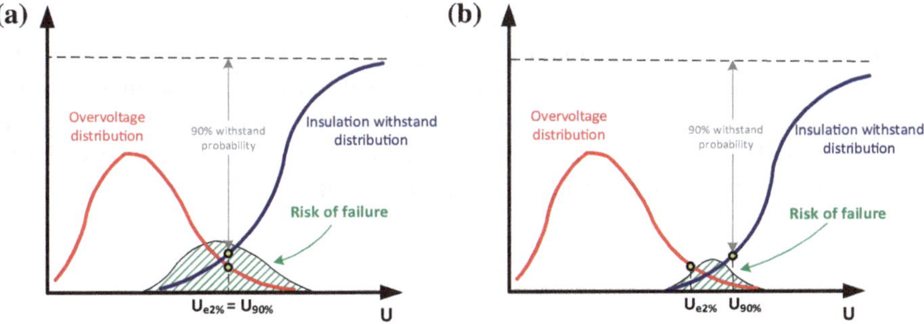

Fig. 3.4 Risk of failure based on the simplified statistical approach **a** $U_{e2\%} = U_{90\%}$ $\gamma = 1$, **b** $U_{e2\%} < U_{90\%}$, $\gamma > 1$

However, the insulation level should be chosen in a way that at least the 2% overvoltage probability ($U_{e2\%}$) coincides with the 90% withstand probability ($U_{90\%}$) as shown in Fig. 3.4a [6]. This corresponds to a statistical safety factor[5] of 1.

The statistical safety factor (γ) is the ratio of statistical withstand voltage to statistical overvoltage [17]. A larger value than 1 for statistical safety factor indicates the higher value of statistical withstand voltage with respect to statistical overvoltage which means that insulation withstand distribution is shifted along the U-axis in Fig. 3.4b. Increasing statistical safety factor reduces the risk of failure, but at the same time increases insulation strength level, and therefore, insulation costs [17]. BS-EN 50341-1 suggests a statistical safety factor of 1.05 in the procedure of insulation coordination for the determination of air clearances on overhead transmission lines. This value corresponds to the risk of failure of 1.0×10^{-3} [4].

[5]The statistical safety factor is equivalent to the statistical coordination factor (Kcs) in IEC Std. 60071-2.

3.3 Insulation Coordination Procedure

Insulation coordination study is needed to determine minimum required air clearances on an overhead transmission line. BS-EN 50341-1 defines different types of air clearances, as given in Table 3.2, which are dependent on the type of voltage or overvoltages. These air clearances are necessary to prevent disruptive discharges between live parts as well as live parts and grounded parts. To derive minimum required air clearances, an insulation coordination procedure is presented in BS-EN 50341-1 [4] which is summarized within a flowchart as shown in Fig. 3.5.

This procedure starts by adopting an appropriate statistical withstand voltage ($U_{90\%\_ff}$), statistical overvoltage ($U_{e2\%\_sf}$) and the highest system voltage (U_s). It is followed by considering additional factors including statistical coordination factor (K_{cs}), atmospheric factor (K_a), deviation factor (K_z), and gap factor (K_g). In the flowchart, f(d) is the withstand voltage of air gap as a function of clearance distance of the air gap (d). To determine the required air clearances for the air gaps, BS EN Standard 50341-1 presents formulas to compute the required minimum phase-to-earth and phase-to-phase air clearances on overhead lines. These formulas together with their parameters, the definitions and relations of the factors are discussed in more details in [11]. The outcome of insulation coordination study is the calculation of minimum required air clearances for different voltage and overvoltage stresses given in Table 3.2.

3.4 Determination of Minimum Required Air Clearances

The insulation coordination procedure shown in Fig. 3.5 has been implemented and consequently, minimum required air clearances based on different voltage stresses (Fig. 3.6) have been calculated for the fully composite pylon. Firstly, statistical withstand voltage and statistical overvoltage are chosen from [2] and a statistical coordination factor of 1.05 is applied to statistical overvoltage. Then, representative

Table 3.2 Definition of different air clearances based on type of voltages and overvoltages

Type of voltage or overvoltage	Minimum required air clearances to avoid a flashover between	
	Phase conductors and other objects at ground potential	Phase to phase conductors
Fast-front overvoltages (ff)	$D_{el\ (ff)}$	$D_{pp\ (ff)}$
Slow-front overvoltages (sf)	$D_{el\ (sf)}$	$D_{pp\ (sf)}$
Power frequency voltage (50 Hz)	$D_{el\ (50\ Hz)}$	$D_{pp\ (50\ Hz)}$

D_{el} is the air clearance between phase conductor and shield wire or tower structure (internal clearance) or is the air clearance between phase conductor and obstacle (external clearance)
D_{pp} is the air clearance between phase to phase conductors (internal clearance)

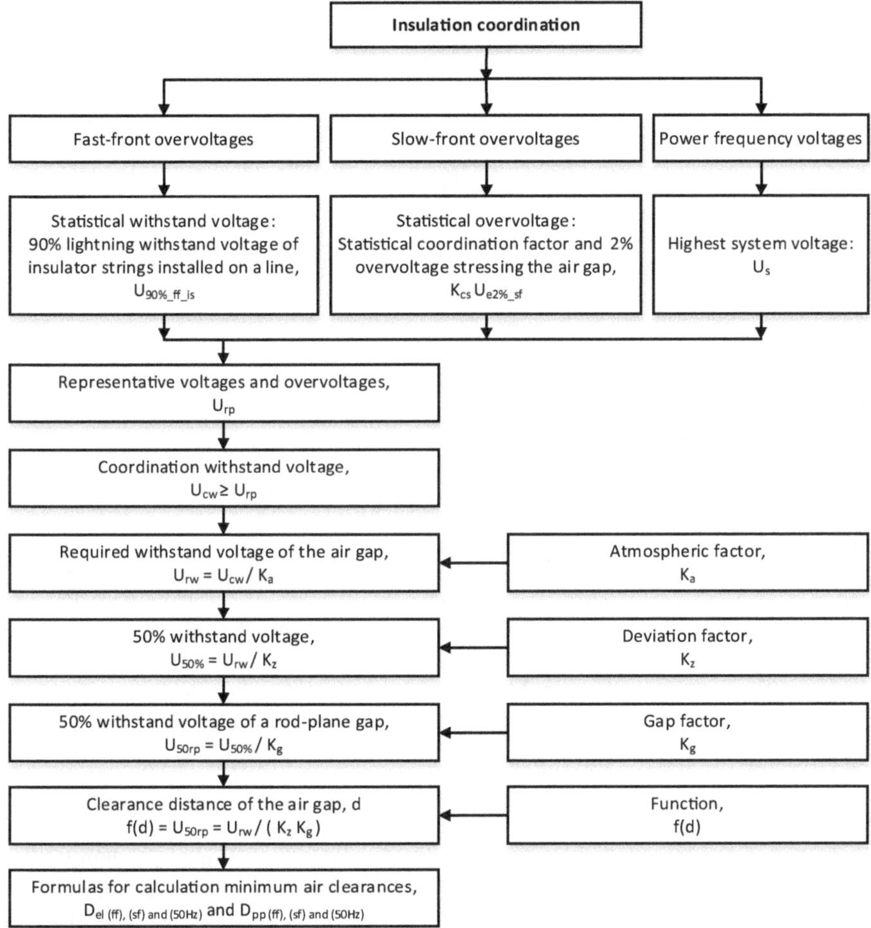

Fig. 3.5 Flowchart of insulation coordination procedure to derive minimum required air clearances [4]

overvoltages for the phase-to-phase and phase-to-shield wire air clearances are determined, and therefore, the required withstand voltages for the air gaps between phase-to-phase and phase-to-shield wire are calculated by applying atmospheric factor, deviation factor and gap factor. The relevant gap factor for the fully composite pylon is conductor-conductor gap configuration.

Afterwards, the required air clearances on the unibody cross-arm between phase-to-phase and phase-to-shield wire are calculated for different voltage stresses of fast-front, slow-front and power frequency voltages. The calculated air clearances on the fully composite pylon are given in Table 3.3. Comparing the three values of D_{el} in Table 3.3 shows that the highest air clearance between top phase conductor

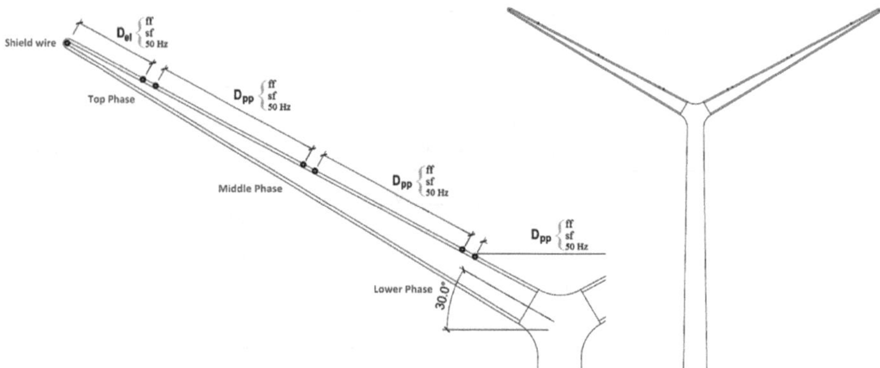

Fig. 3.6 Air clearances based on different voltage stresses on the unibody cross-arm of fully composite pylon

Table 3.3 Calculated air clearances on the fully composite pylon [11]

Type of air clearance	Calculated air clearance between	
	Top phase conductor and shield wire K_g (Conductor-Conductor) K_a (1000 m)	Phase to phase conductors K_g (Conductor-Conductor) K_a (1000 m)
Fast-front (ff)	$\mathbf{D_{el\,(ff)} = 2.69\ m}$	$D_{pp\,(ff)} = 3.23$ m
Slow-front (sf)	$D_{el\,(sf)} = 2.23$ m	$\mathbf{D_{pp\,(sf)} = 3.68\ m}$
Power-frequency (50 Hz)	$D_{el\,(50\,Hz)} = 0.68$ m	$D_{pp\,(50\,Hz)} = 1.17$ m

and shield wire is 2.69 m, which is corresponds to the fast-front overvoltages. This means that the lightning impulse voltages have determinant role in the allocation of air clearance between live parts and grounded parts. $D_{el\,(50\,Hz)}$ is meaningless for the fully composite pylon, because all phase conductors are fixed on the unibody cross-arm by conductor clamps and there is no suspension insulator on the pylon. This parameter is usually used in conventional steel lattice tower designs to consider the effect of wind on the deviation angle of suspension insulator sets at the tower top.

The highest value of D_{pp} in Table 3.3 is 3.68 m and is related to the slow-front overvoltages (switching impulse voltages). $D_{pp\,(50\,Hz)}$ implies to the air clearance between two phases' conductors in the mid-span, which should be enough to prevent flashover between the adjacent phases due to wind action or galloping. The highest values of 2.69 and 3.68 m at the tower top are chosen as the calculated air clearances between top phase conductor and shield wire, as well as between phase-to-phase conductors, respectively. These values have been computed based on the adopted Lightning Impulse Withstand Voltage (LIWV) of 1550 kV and Switching Impulse Withstand Voltage (SIWV) of 1050 kV for the fully composite pylon [11].

The recommended values of air clearances by CIGRE, IEC and BS standards for the highest system voltage of 420 kV have been presented in [11] and compared with the calculated values. Table 3.4 shows standard sets of fast-front and slow front impulse voltages for the highest system voltage of 420 kV. Non-standard set of impulse voltages for the fully composite pylon is due to the adopted Lightning Impulse Withstand Voltage (LIWV) of 1550 kV, which is associated with the highest system voltage of 550 kV. The reason is that the unibody cross-arm of fully composite pylon has different insulation configuration with respect to other conventional insulators, which makes it difficult to repair or replace the cross-arm's components.

The standard value of 1050 kV has been adopted for the Switching Impulse Withstand Voltage (SIWV) of fully composite pylon, because of utilizing surge arresters and other protective devices in both ends of the line. The application of surge arresters can effectively restrict the appeared overvoltages in the system below the withstand voltage of equipment. Although in high voltage power system, the surge arresters are used to protect the equipment of substations, they also can help to avoid oversizing unibody cross-arm of fully composite pylon. This is due to the fact that surge arresters can prevent the penetration of switching overvoltages into the overhead lines and can restrict the possible overvoltages caused by switching operations in substations. Therefore, a reduced Switching Impulse Withstand Voltage (SIWV) can be selected for the air clearances of unibody cross-arm.

Based on Table 3.4, different values for D_{el} have been reported in the IEC Std. 60071-1 [2], IEC Std. 60071-2 [3], BS-EN 50341-1 [4] and empirical guidelines in CIGRÉ TB 72 [14]. The discrepancy in the recommended air clearances for D_{el} is mainly due to the difference in the considered gap configurations. It can be observed from Table 3.4 that the D_{pp} with the common conductor-conductor gap configuration has approximately a constant value for each set of impulse voltages. The recommended values for D_{pp} have good agreement with the calculated one and therefore the D_{pp} of 3.68 m is chosen as the minimum required phase-to-phase air clearance on the fully composite pylon. In the case of D_{el}, the values of 2.9 and 3.02 m are derived from IEC Std. 60071-1, 2 and BS-EN 50341-1, respectively, for the non-standard set of impulse voltages. These values are higher than the calculated value of 2.69 m, which is due to their lower values of gap factors in comparison to the conductor-conductor gap factor of 1.6. On the other hand, CIGRÉ TB 72 proposes three different values for D_{el} that are independent from the gap configuration. However, a D_{el} of 2.80 m is placed among these three values and is a commonly used European value for the external clearance at 420 kV [4]. Since this value has been demonstrated to be enough for the safety of general public, therefore, instead of the calculated value of 2.69, the D_{el} of 2.80 m has been chosen in [11] as a conservative value for the minimum required phase-to-earth air clearance on the fully composite pylon.

Table 3.4 Recommended and calculated air clearances under impulse voltages [11]

Highest system voltage	Impulse voltages		IEC 60071-1, 2		EN 50341-1		CIGRÉ TB 72		Calculated values (based on EN 50341-1)	
			Up to 1000 m		K_a (1000 m)				K_a (1000 m)	
U_s	$U_{90\%\text{-ff}}$	U_{sf}	$\mathbf{D_{pe}^{a}}$	$\mathbf{D_{pp}}$	$\mathbf{D_{el}}$	$\mathbf{D_{pp}}$	$\mathbf{D_{pe}^{a}}$	$\mathbf{D_{pp}}$	$\mathbf{D_{el}}$	$\mathbf{D_{pp}}$
			$K_g \approx 1.45$ (Cond.-Struc.)	$K_g \approx 1.6$ (Cond.-Cond.)	$K_g = 1.3$ (Cond.-Obst.)	$K_g = 1.6$ (Cond.-Cond.)			$K_g = 1.6$ (Cond.-Cond.)	
kV	kV	kV	m	m	m	m	m	m	m	m
420[b]	1175	850	2.2 (LIWL)	2.6 (SIWL)	2.25 (SIWL)	2.67 (SIWL)	2.40	2.70	–	–
	1300	950	2.4 (LIWL)	3.1 (SIWL)	2.64 (SIWL)	3.15 (SIWL)	**2.80**	3.20	–	–
	1425	1050	2.6 (LI/SIWL)	3.6 (SIWL)	3.02 (SIWL)	3.68 (SIWL)	3.25	**3.73**	–	–
420[c]	**1550**	**1050**	**2.9** (LIWL)	**3.6** (SIWL)	**3.02** (SIWL)	**3.68** (SIWL)	–	–	**2.69** (LIWL)	**3.68** (SIWL)

[a] D_{pe} is the same as D_{el} and refers to the phase-to-earth clearances
[b] Standard set of impulse voltages for traditional towers
[c] A non-standard set of impulse voltages for the unibody cross-arm of fully composite pylon
Lightning Impulse Withstand Voltage (LIWV) and Switching Impulse Withstand Voltage (SIWV) imply to the determinant insulation level
LIWL and SIWL are equivalent to basic lightning impulse insulation level (BIL) and basic switching impulse insulation level (BSL), respectively

3.4.1 Internal and External Clearances at the Tower Top and Mid-Span

Based on selected phase-to-earth and phase-to-phase air clearances, internal clearances at the top of fully composite pylon and mid-span are specified using the guidelines of National Normative Aspects for Denmark (NNAs—DK) reported in BS-EN 50341-3 [5]. Minimum required internal clearances have been determined for 3 load cases reported in NNAs—DK including Maximum Design Temperature (MDT), Conductor Swing (CS) and Galloping conductor (GC). Figure 3.7 illustrates the internal clearances at the tower top and mid-span for different load cases. Since there are no suspension insulators at the top of fully composite pylon and the phase conductors are fixed using conductors clamps on the unibody cross-arm, the air clearances at the top of the pylon are identical for three load cases of Maximum Design Temperature (MDT), Conductor Swing (CS) and Galloping conductor (GC).

Regarding air clearances in the mid-span, under the condition of extreme swinging of conductors, the probability of occurrence of a lightning or switching surge is very small and negligible [16]. For this reason, lower air clearances are dedicated in the mid-span for the load cases of Conductor Swing (CS) and Galloping conductor (GC).

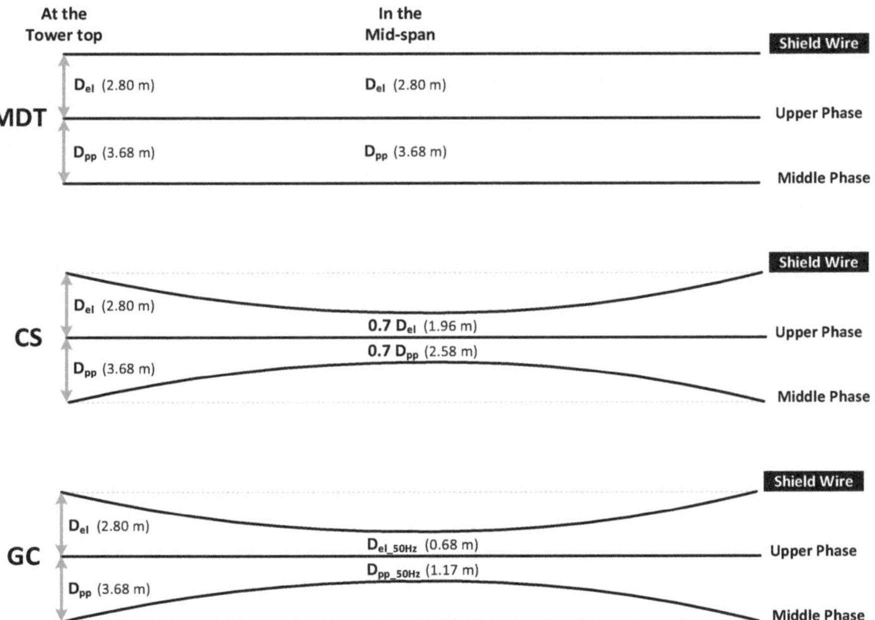

Fig. 3.7 Internal clearances along a whole span at the tower top and mid-span for different load cases including Maximum Design Temperature (MDT), Conductor Swing (CS) and Galloping conductor (GC)

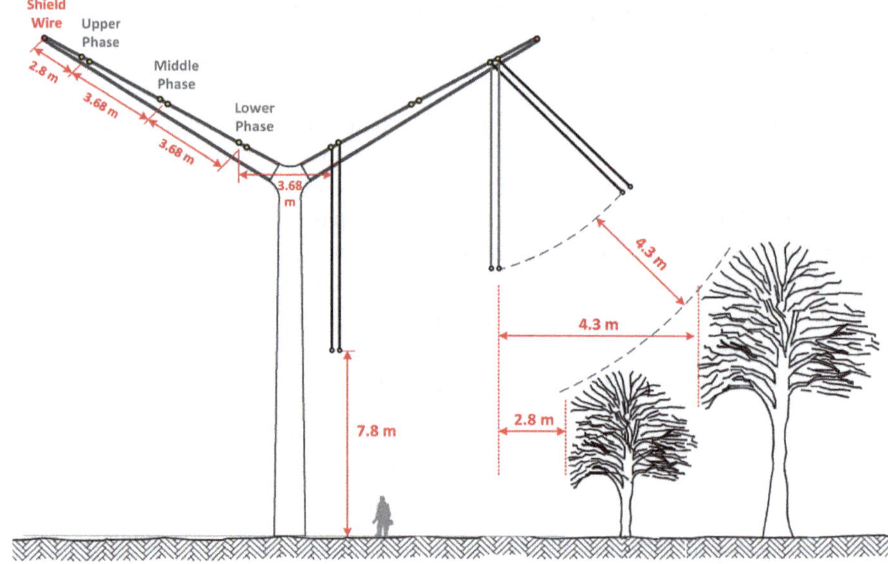

Fig. 3.8 Internal air clearances on the unibody cross-arm of fully composite pylon (left side) and external clearances to ground in the areas remote from buildings and roads (right side)

Finally, the determination of internal clearances leads to the basic dimensioning of fully composite pylon. The overall length of unibody cross-arm can now be specified for the mechanical design of pylon and the sag of conductors in the mid-span can be determined by considering the assigned height for the fully composite pylon and minimum required external air clearances in the mid-span. For the case of external clearances, BS-EN 50341-3, NNAs—DK [5] gives some guidelines regarding minimum required external clearances, which are shown in Fig. 3.8. These clearances are necessary to prevent danger to general public and those working adjacent to the overhead line.

3.5 Summary

In this chapter, the deterministic approach of insulation coordination study is briefly described. The shapes and durations of power frequency, slow-front and fast-front overvoltages are mentioned. Statistical behavior and failure risk of air insulation are implied. The influence of statistical withstand voltage and statistical overvoltage on the risk of failure is depicted which is one of the important parameters in the determination of air clearances of overhead transmission line pylons.

The insulation coordination procedure to determine required electrical clearances on the fully composite pylon is presented. The calculated phase-to-earth and phase-to-phase air clearances are mentioned and have been computed based on the guidelines of different standards and literature sources. The calculated values are compared with the recommended values of standards and consequently, a conservative value and a calculated value are adopted for the phase-to-earth and phase-to-phase air clearances, respectively. The internal and external air clearances at the top of the pylon and within the mid-span are derived based on the adopted air clearances. As a result, the basic dimension of fully composite pylon is estimated in terms of electrical point of view.

References

1. National Electrical Safety Code 2007 Edition, IEEE Standard C2-2007
2. IEC 60071-1, Insulation co-ordination—Part 1: definitions, principles and rules (2011)
3. IEC 60071-2, Insulation co-ordination—Part 2: application guide (1996)
4. BS EN 50341-1, Overhead electrical lines exceeding AC 1 kV—Part 1: general requirements—-Common specifications, British Standards (2012)
5. BS EN 50341-3, Overhead electrical lines exceeding AC 45 kV—Part 3: set of National Normative Aspects, British Standards (2001)
6. R. Bhattarai, *Uprating of Overhead Lines* (Cardiff University, 2011)
7. IEEE Std C62.82.1, IEEE Standard for Insulation Coordination—Definitions, Principles, and Rules (2010)
8. IEEE Std 1313.2, IEEE Guide for the Application of Insulation Coordination (1999)
9. IEEE Std 1427, IEEE Guide for Recommended Electrical Clearances and Insulation Levels in Air-Insulated Electrical Power Substations (2006)
10. M.C.V. Rosário, Coordenação de isolamento em linhas aéreas Metodologias e Aplicação à Compactação de Linhas REN (Technical University of Lisbon, 2011)
11. T. Jahangiri, C.L. Bak, F.F. da Silva, B. Endahl, Determination of minimum air clearances for a 420 kV novel unibody composite cross-arm, in *2015 50th International Universities Power Engineering Conference (UPEC)* (2015), pp. 1–6
12. S. Venkatesan, R. Bhattarai, M. Osborne, A. Haddad, H. Griffiths, N. Harid, A case study on voltage uprating of overhead lines—air clearance requirements, in *45th International Conference on Universities Power Engineering Conference (UPEC)* (2010)
13. T. Jahangiri, Electrical Design of a New, Innovative Overhead Line Transmission Tower Made in Composite Materials (Aalborg University, 2018)
14. K.O. Papailiou, *CIGRE Green Book: Overhead Lines* (Springer, 2017)
15. IEC 60071-4, Insulation co-ordination—Part 4: computational guide to insulation co-ordination and modelling of electrical networks (2004)
16. F. Kiessling, P. Nefzger, J.F. Nolasco, U. Kaintzyk, *Overhead Power Lines, Planning, Design, Construction* (Springer, Heidelberg, 2003)
17. E. Kuffel, W.S. Zaengl, J. Kuffel, *High Voltage Engineering Fundamentals*, 2nd edn. (Butterworth-Heinemann, 2000)

Chapter 4
Electrical Design of Fully Composite Pylon

4.1 Introduction

An efficient design of insulation for the unibody cross-arm of fully composite pylon is one of the major challenges in the pylon design, which should be done by considering existing literatures and experiences for the design of HV composite cross-arms. In this chapter, the recommendations and guidelines of international standards related to the design of shed profiles are presented, in order to design a proper shed profile for the insulation of unibody cross-arm. By adopting a well-designed shed profile, the electrical performance of entire insulation on the unibody cross-arm is evaluated. Some design criteria and issues, which should be addressed and utilized in the evaluation of electric field performance of fully composite pylon, are presented. Two design scenarios of utilizing internal and external ground connection for the shield wires are examined by performing numerous finite element analyses. Electric field magnitudes in the different regions of interest on the unibody cross-arm are computed and compared with the electric field criteria. Based on the assessment of electric field results obtained by finite element analyses, it has been found that utilization of corona rings to modify electric field magnitudes around and inside the unibody cross-arm is inevitable. Therefore, the sizes and positions of corona rings are optimized to satisfy electric field magnitudes in the critical parts on the unibody cross-arm.

4.2 Insulation Design

4.2.1 Creepage Distance

Creepage distance is the shortest distance along the insulation surface as shown in Fig. 4.1. Operating voltage is the basic parameter to determine the creepage distance

T. Jahangiri et al., *Electrical Design of a 400 kV Composite Tower*, Lecture Notes in Electrical Engineering 557,
https://doi.org/10.1007/978-3-030-17843-7_4

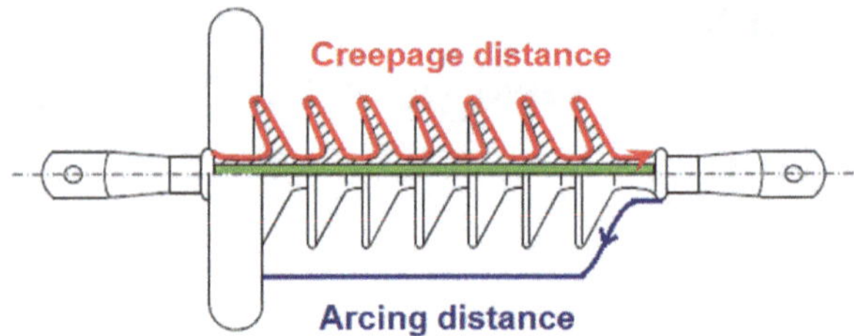

Fig. 4.1 Creepage distance and arcing distance of an insulator

along insulation whereas, the effect of switching and lightning impulses are taken into account to determine arcing distance at the air, which was discussed in the previous chapter. The unibody cross-arm of fully composite pylon has two different creepage distances along its insulation. One is the creepage distance between two adjacent phases whereas another is between the upper phase and shield wire.

Traditionally, the dimensioning of insulation for outdoor use is widely based on creepage distance and can be done according to the "only" minimum requirements reported in IEC 60815-3 [1, 2]. Outdated version of IEC Standard 60815 (1986) [3] defined 'Specific Creepage Distance (SCD)' for different pollution levels to calculate minimum nominal creepage distance between phase to phase. Whereas, an updated version of IEC standard 60815 (2008) [4–6] uses 'Unified Specific Creepage Distance (USCD)' to estimate minimum required creepage distance between phase to ground. However, in order to determine minimum required creepage distances on the unibody cross-arm, the newer term 'Unified Specific Creepage Distance (USCD)' has been used. USCD definition based on IEC standards 60815-1, 3 [4, 6] is the creepage distance of an insulator divided by the RMS value of the highest operating voltage across the insulator.

Adopting an appropriate USCD for an insulator depends on the severity of pollution at the site where the insulator will be installed. The standard values of USCD for different pollution levels are given in Table 4.1. In Table 4.1, the minimum required creepage distances on the unibody cross-arm are given for different site pollution severity (SPS) classes. As it can be seen in Table 4.1, the very light pollution level needs lower creepage distances among the other pollution levels. The highest values of creepage distances are related to the high pollution level. In this study, the SPS class of medium pollution level is considered for the dimensioning of insulation on the fully composite pylon. Because, the moderate pollution condition was chosen at the first stage of the pylon design to calculate required creepage distances on the unibody cross-arm.

Required creepage distances on the unibody cross-arm between the phase-to-phase and phase-to-shield wire regions should be provided by sheds. The number of sheds in the regions depend on shed profiles and corresponding creepage and arcing distances.

Table 4.1 Minimum required creepage distances on the unibody cross-arm based on different site pollution severity

Site pollution severity (Pollution level)	Unified specific creepage distance (IEC/TS 60815-3) (USCD)	Minimum required creepage distance in the region of	
		Phase-to-shield wire	Phase-to-phase
	(mm/$kV_{ph\text{-}ground}$)	(mm)	(mm)
a—Very light	22.0	5335	9240
b—Light	27.8	6741	11676
c—Medium	**34.7**	**8414**	**14574**
d—Heavy	43.3	10500	18186
e—Very heavy	53.7	13021	22554

4.2.2 Shed Profile

Conventional composite insulators' designs differ in many aspects especially in the configuration of sheds, which have a high influence on the contamination performance of the insulators [7]. Although composite insulators have good pollution performance in comparison to other porcelain and glass insulators, their pollution performance depend on the design of shed profiles which should be appropriately designed for the use in situ condition. The proper design of shed profile is important to [1, 6]:

- avoid pollution traps,
- avoid rain bridging,
- aid self-cleaning,
- prevent local short-circuiting between sheds,
- control local electric field stress.

Some general recommendations are presented in [4–6] to select suitable shed profiles based on the service condition of insulator and dimensioning the sheds to achieve an acceptable pollution performance. Profiles of sheds can be categorized into two categories: open profiles or steep profiles, which differs due to the slope of sheds. Sheds with open profiles, shown in Fig. 4.2, have excellent self-cleaning capability and have the benefit of using in all environments such as coastal areas, desert and heavily polluted industrial areas [6]. The open profiles can be used in horizontal and vertical orientations and they have a top slope of 20° or less and their underside slope is similar or less. Sheds with steep profiles (anti-fog profiles) are more suitable for the areas where exposed to liquid pollutants such as acid rain or salt water.

The majority of shed profile designs for composite insulators are in the form of uniform and alternating sheds which are illustrated in Fig. 4.3. Utilization of alternating profile with large and small sheds are also more common in the design of equipment for transmission lines. The alternating shed profiles have the benefit of

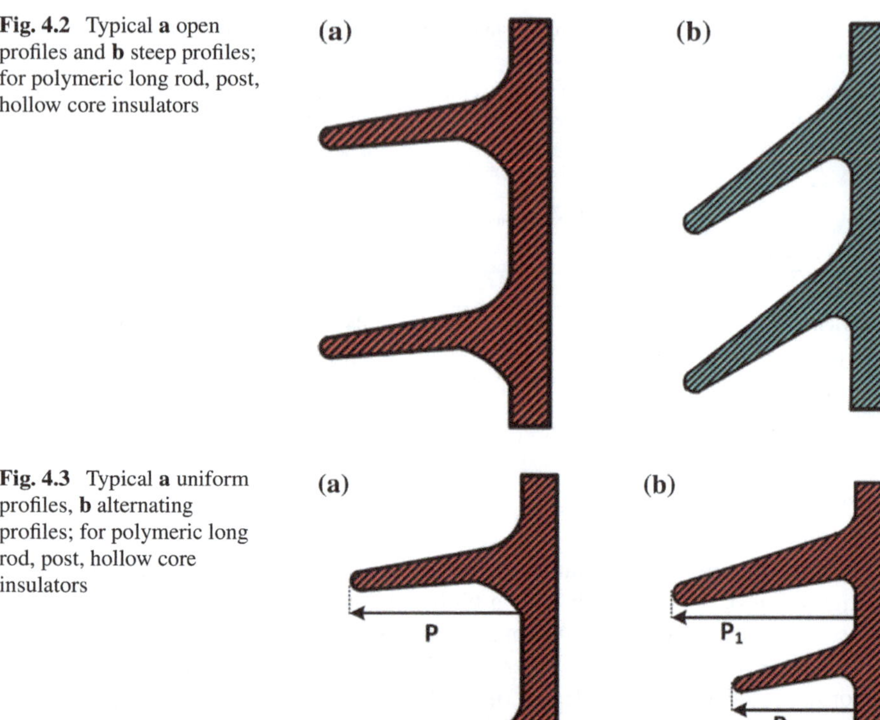

Fig. 4.2 Typical **a** open profiles and **b** steep profiles; for polymeric long rod, post, hollow core insulators

Fig. 4.3 Typical **a** uniform profiles, **b** alternating profiles; for polymeric long rod, post, hollow core insulators

increased creepage distance per unit length and, at the same time, can provide good wet performance in heavy rain (or icing) and good self-cleaning capability.

Based on IEC/TS 60815-3, an insulator's shed profile can be classified into the following parameters [1, 6] which are illustrated in Fig. 4.4:

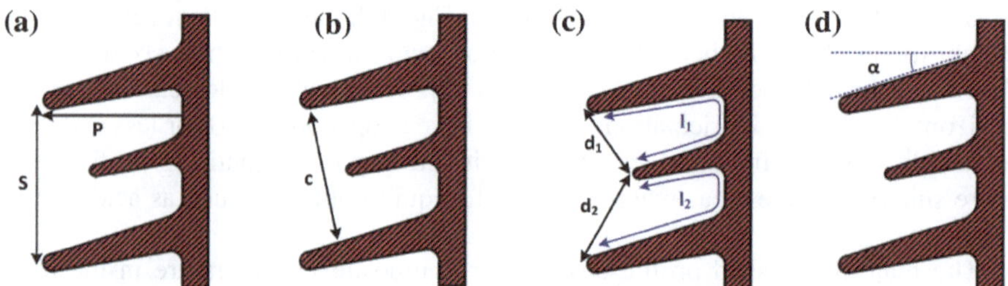

Fig. 4.4 **a** Spacing versus shed projection, **b** Minimum distance between sheds, **c** Creepage distance versus clearance, **d** Shed's top surface inclination angle

- Spacing versus shed projection (s/p),
- Minimum distance between sheds (c),
- Creepage distance versus clearance (l/d),
- Shed angle (α),
- And creepage factor for the entire of insulator (CF = l/A),

All parameters of (s/p), (c), (l/d) and (α) involve shed to shed spacing and therefore, are important to avoid shorting out creepage path bridged by shed to shed arc [6]. For example, shed to shed arcing can occur during icing event or ice shedding conditions when electric field stress between air gaps increases and causes bridges with low energy arcs. In the case of development of arcs along the insulator, critical flashover voltage reduces and a flashover can occur across the entire of insulator. The arc development paths and flashover voltages depend on the dimensions of shed profiles [8]. Alternating profiles with higher shed to shed spacing or with three or four different shed projections [9] can have better performance during average and heavy icing conditions whereas, insulators with bigger diameters or with vertical orientation seem to perform worse under ice and snow conditions [10]. Conversely, insulators which are installed horizontally or installed with an angle have better behavior [10]. On the other hand, contamination flashover voltage of composite insulators has a relationship with the parameters of shed profiles. In overall, the effect of shed parameters on the contamination flashover voltage can be summarized as follows [7]:

Increasing shed projection causes:

- Increasing in the creepage distance of composite insulator (positive for pollution flashover performance)
- Decreasing in the utilization rate of creepage distance (negative for economy efficiency)

Decreasing shed projection causes:

- Promotion of contamination characteristics of composite insulator (negative for pollution flashover performance)
- Increasing in the utilization rate of creepage distance (positive for economy efficiency)

Increasing shed spacing causes:

- Preventing bridge down between sheds in raining or ice coating weather (positive for pollution flashover performance)
- Promotion in the utilization rate of creepage distance (positive for economy efficiency)

Decreasing shed spacing causes:

- Increasing in the creepage distance of composite insulator if the insulation space is definite (flashover voltage would not grow linearly with the creepage distance)
- The arc can easily bridge two sheds (if shed spacing is too small)

Table 4.2 Recommended ranges of profile parameters based on IEC/TS 60815-3

Parameter	Dimension/ Position/ SPS class	Shed profile	Deviation		
			Major	Minor	None
s/p	Shank diameter > 110 mm	Without under ribs	0.4–0.5	0.5–0.65	0.65–1
c (mm)	All	Uniform	20–22.5	22.5–25	25–50
		Alternating	20–30	30–40	40–50
l/d	All	All	5.5–7	4.5–5.5	0–4.5
α	Vertical	All	35°–60°	0°–5° 25°–35°	5°–25°
	Horizontal		30°–60°	20°–30°	0°–20°
CF	SPS class c	All	5–5.5	4.5–5	2.5–4.5

On the other hand, the last parameter of shed profile is creepage factor for the entire of insulator. The creepage factor (CF) or creepage distance density is the ratio of total creepage distance of an insulator over the arcing distance of the insulator. By using creepage factor, the overall density of assigned creepage distance (creepage distance per unit length) can be checked and it can be also indicated how much the sheds are tightly arranged. Usually, the parameter of creepage factor can be automatically realized if the requirements of shed projections, spacing vs projection and minimum distance between sheds are met [6].

Based on how much the aforementioned parameters can reduce the pollution performance of the insulator, the values of parameters are classified in the three classes: none deviation (the profile parameter is within normal range), minor deviation (the profile parameter can lead to reduction in pollution performance) and major deviation (the profile parameter can seriously affect the pollution performance of insulator) [1].

The IEC/TS 60815-3 recommended that only one deviation in 'minor deviation' class is allowable, and a parameter in 'major deviation' is not acceptable [1, 10]. In the cases with multiple 'minor deviation' or one 'major deviation', it is mentioned that the design parameters should be changed to resolve the problem. The recommended ranges of the shed parameters based on their applicability for the unibody cross-arm of fully composite pylon are sum upped and given in Table 4.2.

For the purpose of the unibody cross-arm, two different shed profile's dimensions are assumed for the insulation of unibody cross-arm as shown in Fig. 4.5. Shed profiles with higher shed spacing are considered for the insulation between upper phase and shield wire (Fig. 4.5a) whereas the same shed profiles with shorter shed spacing are assigned for the insulation between two adjacent phases (Fig. 4.5b).

Although the shed spacing depends on the dimensions of shed profiles, it is mainly dependent on the required creepage distance in the region and the corresponding arcing distance. Phase-to-shield wire region has a 2.80 m arcing distance and its required creepage distance is lower than that for the phase-to-phase region with a

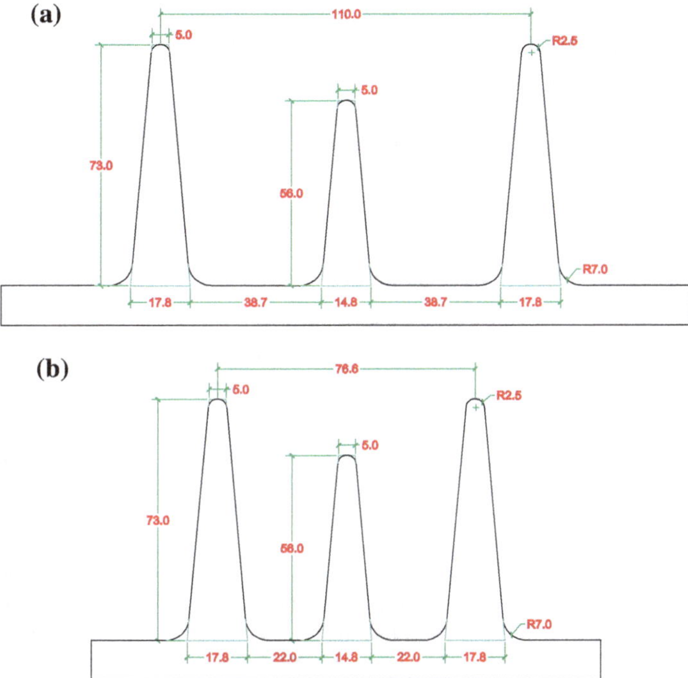

Fig. 4.5 Shed profiles' dimensions in the region of **a** phase to shield wire, **b** phase to phase

3.70 m arcing distance. Therefore, the sheds in the phase-to-phase region would be placed tightly in comparison to the phase-to-shield wire region.

The values of large and small shed projections are adopted from standard profile designs reported in [11]. Moreover, a tapered shape for the shed profiles is chosen because it gives better flexibility for sheds to avoid damage during transportation and installation and also to facilitate the easier release of insulation from the mould during injection moulding process [10]. The tip of sheds is rounded with a radii of 2.5 mm to minimize electric field magnitudes on tips and to avoid from potentially occurrence of corona discharges. In the same manner, the root of sheds is also filleted to manage electric field magnitudes in the interface between shed and sheath.

However, in order to verify the suitability of shed profiles in both regions of phase-to-shield wire and phase-to-phase, their parameters are given in Table 4.3 and evaluated with the recommended values by IEC/TS 60815-3 in Table 4.2. According to Table 4.3, all shed profile parameters are in None deviation class and therefore comply with all the requirements mentioned in Table 4.2. Total creepage distances in the regions of phase-to-phase and phase-to-shield wire are 6.8% and 6.7%, respectively, higher than the minimum required creepage distances specified in Table 4.1. When the diameter of an insulator is bigger than 300 mm, increasing in total creepage distance is desirable [10], hence, the values of total creepage distances in both regions

Table 4.3 Evaluation of shed parameters in two regions of phase-to-shield wire (ph-gnd) and phase-to-phase (ph-ph) based on deviation classes in IEC/TS 60815-3

Alternating large and small shed	Parameter	Deviation
	$p_1 = 73$ mm $p_2 = 56$ mm	
	$p_1 - p_2 = 17$ mm	None
	Number of sheds: $(52^{\text{Large}} + 52^{\text{Small}})_{\text{ph-ph}}$ Number of sheds: $(27^{\text{Large}} + 27^{\text{Small}})_{\text{ph-gnd}}$	
	$(s/p)_{\text{ph-ph}} = 76.6$ mm/73 mm $= 1.05$ $(s/p)_{\text{ph-gnd}} = 110$ mm/73 mm $= 1.50$	None None
	$(c)_{\text{ph-ph}} = 71$ mm $(c)_{\text{ph-gnd}} = 105$ mm	None None
	$(l/d)_{\text{ph-ph}} = 299.3$ mm/83.8 mm $= 3.57$ $(l/d)_{\text{ph-gnd}} = 332.7$ mm/115.1 mm $= 2.89$	None None
	$(\alpha) = 5°$	None
Creepage distance	$(L)_{\text{ph-ph}} = 15567$ mm > min 14574 mm $(L)_{\text{ph-gnd}} = 8984$ mm > min 8414 mm	(6.8% higher) (6.7% higher)
Arcing distance	$(A)_{\text{ph-ph}} = 3700$ mm (Between two corona rings) $(A)_{\text{ph-gnd}} = 2800$ mm (Between corona ring and shield wire)	
Creepage factor	$(CF)_{\text{ph-ph}} = 15567$ mm/3700 mm $= 4.20$ $(CF)_{\text{ph-gnd}} = 8984$ mm/2800 mm $= 3.20$	None None

are reasonable, because the overall longitudinal cross-section of unibody cross-arm is much larger than 300 mm. Therefore, Table 4.3 indicates that these shed profile designs will provide good pollution performance and self-cleaning capability for the insulation of fully composite pylon.

4.3 Electric Field Considerations

To design composite insulators especially for extra high voltage (EHV) overhead lines, electric field considerations have a great importance and should be considered particularly. High electric field intensities around composite insulators can cause corona discharges on their surfaces. Corona discharge is a low power discharge, which has audible and luminous aspects and can be observed especially in the sharp edges of a high voltage electrode where inhomogeneous high electric fields exist.

In an AC overhead transmission line, high electric field magnitudes are on the surfaces of phase conductors and metallic hardwares of insulators especially near to the energized ends. The strong electric field on the metallic parts and conductors leads to corona discharge, which is responsible for corona losses, audible noise and radio interference. Corona discharge is also responsible for the generation of ozone, light emission and corona induced vibrations in the conductor [10].

The electric field distribution on composite insulators depends on the voltage of line, configuration of insulator, tower structure, dielectric material properties, position and dimensions of corona rings, and the shape of end fittings. High electric field magnitudes within and on the surface of composite insulators can affect their short and long-term performances. For example under wet condition, electric field magnitudes higher than water droplet corona onset can lead to enhancement in corona activities on the insulator's surface and accelerate the aging of its material [12, 13]. Water droplet corona discharge usually occurs between droplets and destroys the hydrophobicity of shed housing. Prolonged exposure to corona can result in permanent loss of hydrophobicity of shed housing and development of cracks on its surface.

A contaminated and wet insulator can provide higher flowing of leakage current on its surface and therefore can cause dry band arc formation, which can lead to a flashover along the insulator. In the regions with high electric filed magnitudes especially near the end fittings, the presence of manufacturing defects such as air bubbles or elongated cavities within the core or in the interface between shed housing and core can result internal partial discharges inside the voids. By developing partial discharges in the voids, organic acids can be produced inside the fiberglass layer. The acids in combination with a mechanical load can cause the corrosion of fiberglass material and subsequently can lead to brittle fracture and mechanical failure of insulator [10].

In the case of utilizing grading rings, more uneven electric field distribution due to improper positioning and dimensioning of corona rings near the end fittings can erode shed housing in the triple junction where metal fitting, air and sheath (shed housing) meet [14, 15]. By erosion of shed housing in the triple junction, the fiberglass core can be exposed to acid attacks [13]. The sealing material (watertight) in triple junction can also be degraded due to higher electric field magnitudes in the region. Therefore, in the design of a composite insulator, electric field distribution around and inside the insulator should be carefully assessed and controlled to prevent or reduce undesirable effects of corona discharges on different parts of the insulator. The main regions on the composite insulator which should be checked in terms of electric field distribution are [16]:

- On the surface of sheds and around the seal of end fitting
- Within fiberglass core and shed housing and also at the interfaces between metal end fitting, shed housing and core
- On the surface of metallic end fittings and associated corona rings

If electric field magnitudes in any of three regions exceed critical values, corona discharge activities can occur and therefore, the insulator's short term or long term performances may be affected. Some critical values of electric field are mentioned in literatures as electric filed criteria, which should be taken into account during design processes.

4.3.1 Electric Field Criteria

To design a composite insulator, electric field intensities along the insulator should be kept as low as possible to restrict or reduce corona discharge activities on the different regions of insulator. In this regard, some electric field criteria for different parts on an insulator are reported in 'IEEE taskforce on Electric Fields and Composite Insulators' [16] and 'Electric Power Research Institute' (EPRI) [17] which are also specified in technical specifications of National Grid in UK [10]. These electric field criteria are indicated for composite insulators under dry and clean conditions [16, 18] and are also adopted in [13, 19–21] for designing composite cross-arms. The adopted electric field criteria to design the composite cross-arms are as follows:

1. Maximum electric field magnitude on the sheath and sheds measured 0.5 mm from the sheath surface should be less than 0.45 kV_{RMS}/mm [1, 10, 13, 19–21].
2. Maximum electric field magnitude on electric field grading devices and metallic end fittings should be below 1.8 kV_{RMS}/mm [1, 10, 13, 21]. Usually in designs, the value of 2.1 kV_{RMS}/mm is utilized as a reference electric field magnitude [16, 20, 21].
3. Maximum electric field magnitude within dielectric materials including weather sheds and core should be lower than 3 kV_{RMS}/mm [1, 10, 13, 19, 21].
4. Maximum electric field magnitude at a triple junction should be below 0.35 kV_{RMS}/mm [1, 10, 13, 21].

In the above-mentioned criteria, maximum permissible electric field magnitude on sheds' surfaces is not directly specified and only the maximum allowable magnitude at 0.5 mm above the sheath surface is determined to be below 0.45 kV_{RMS}/mm. However, it is mentioned in [14] that the threshold value of water droplet corona onset (on shed housing) is in the range of 0.5–0.7 kV_{RMS}/mm. Practically, keeping the electric field magnitude below this range seems to be not feasible. On the other hand, it is reported in [1, 15] that water droplet corona discharges depend on the hydrophobicity state of shed housing and may occur in the presence of electric field stresses in the range of 0.8–1.3 kV_{RMS}/mm. Therefore, for a reliable design purpose, it is better that maximum electric field magnitude on the surface of weather sheds be kept below 0.8 kV_{RMS}/mm to limit corona discharge activities under wet conditions.

The electric field criteria are useful and relevant for the design of unibody cross-arm of fully composite pylon and they can be used as evaluation factors to check whether local electric filed magnitudes on different parts of unibody cross-arm and related conductor clamps comply with the critical values. A successful design can be achieved by controlling electric field distribution in a way that local electric field magnitudes on different parts of fully composite pylon do not exceed the critical values. Assessment of electric field distribution on the fully composite pylon can be done by using finite element method.

4.4 Finite Element Analysis of Fully Composite Pylon

Nowadays, FEM is a precise and practical method in high voltage engineering and has been widely used for the electrical design of high voltage equipment [20–22]. By recent advancement in high-performance computing, large-scale physical models can be modeled in finite element packages such as ANSYS and Comsol. The modeling of large-scale equipment is usually time-consuming and the solution time depends on the number of finite elements in the entire of model and the available computing power. The applications of finite element electric field analysis for high voltage equipment such as power pylons are presented in [18, 20, 21, 23, 24]. Its application is also reported for large-scale models such as substation [22] and wall bushings [25, 26].

At the initial stage of an equipment design, different equipment modeling can be modeled and evaluated based on FEM results and the variations in design parameters can also be considered to improve the performance of equipment. Subsequently, the results of finite element modeling in final design should be validated by laboratory tests. This design process is an economical approach which eliminates the requirement for numerous prototypes manufacturing [21, 27]. FEM gives better insight into critical design parameters, and therefore, the components' structure can be optimized to achieve satisfactory results before prototype manufacturing.

4.4.1 Basic Design of Fully Composite Pylon

So far, the whole structure of fully composite pylon has not been manufactured in the real world and electric field studies are limited to finite element simulations in full-scale and laboratory tests for a limited phase-to-phase length of the unibody cross-arm. At the first stage of pylon design, an extensive research has been carried out in [1] to get in-depth knowledge regarding the electric field distribution around and inside the fully composite pylon. The configuration and material of conductor clamps were not decided and approved at that time, therefore, the electric field performance of fully composite pylon without conductor clamps was investigated by using finite element modeling for the fully composite pylon with/without internal ground cable through the unibody cross-arm.

In the case of presence of internal ground cable, it was observed in [1] that the electric potential distribution around the unibody cross-arm was restricted to the top

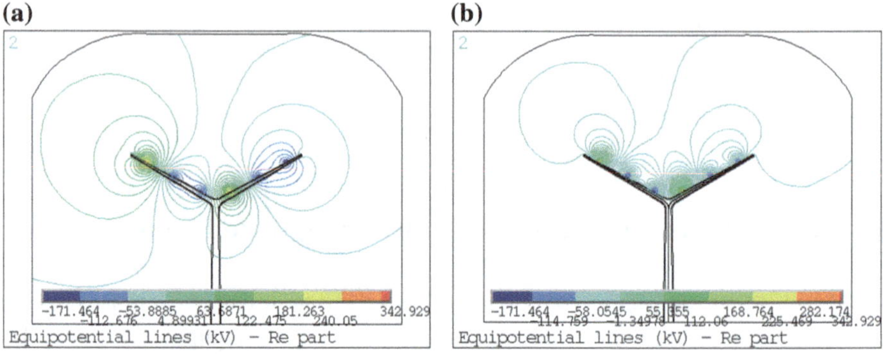

Fig. 4.6 Equipotential lines around fully composite pylon **a** without internal ground cable, **b** with internal ground cable [1]

of fully composite pylon while the electric potential distribution around the pylon without ground cable was more uniform as shown in Fig. 4.6.

The electric field magnitudes around and inside the unibody cross-arm with and without ground cable also showed that the cross-section of unibody cross-arm is under higher electric field stresses when the internal ground cable is passed through its hollow body. The electric field magnitudes on sheds along the unibody cross-arm are illustrated in Fig. 4.7 for two cases of with and without ground cable. Comparing Fig. 4.7a and b shows that utilizing ground cable inside the unibody cross-arm has a serious effect on the electric field magnitudes under the energized regions. It is due to this fact that the electric field stresses on the pylon without ground cable depend on the height of bundled conductors to earth surface whereas the distance between bundled conductors and ground cable has a large impact on the magnitudes of electric field on the unibody cross-arm. However, the region under the top phase of the pylon with ground cable exposes higher electric field intensities which is caused by the conical shape of cross-arm. In the next stage of electric field studies, different non-conductive conductor clamp's designs are added into the pylon's modeling in [21] and are assumed to be attached between phase conductors and the unibody cross-arm. Firstly, a non-conductive conductor clamp design with metal bolts is suggested by [28] which is shown in Fig. 4.8. From finite element analysis results, it was observed that air around steel bolts and top of the clamp were exposed to the breakdown of air insulation. The reason was due to the inducing floating potential on steel bolts that imposed very high electric field strength in the air around the bolts (Fig. 4.9). As a result, the idea of utilizing metal bolts inside the non-conductive conductor clamp was discarded.

A fully non-conductive conductor clamp with a simplified geometry was proposed in [21] as the second scenario in conductor clamp design. As shown in Fig. 4.10, finite element results showed that the electric field magnitudes around phase conductors (shown by gray color around conductors) are still higher than air insulation breakdown and therefore, the height of the conductor clamp should be increased to

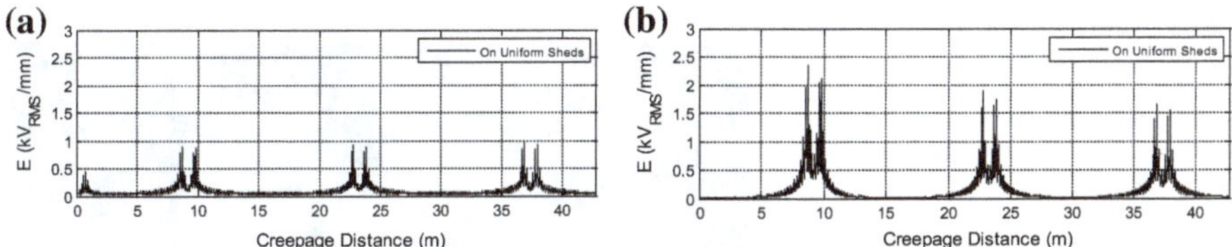

Fig. 4.7 Electric field magnitudes on the sheds of unibody cross-arm **a** without internal ground cable, **b** with internal ground cable [1]

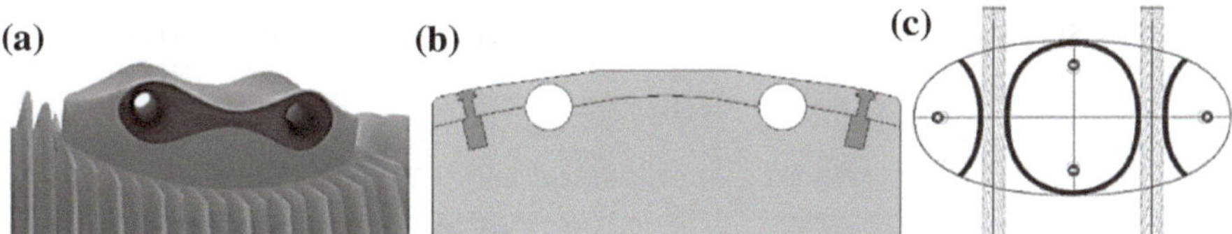

Fig. 4.8 **a** non-conductive conductor clamp on unibody cross-arm, **b** cross-sectional view and **c** top view of the clamp with four steel bolts [21]

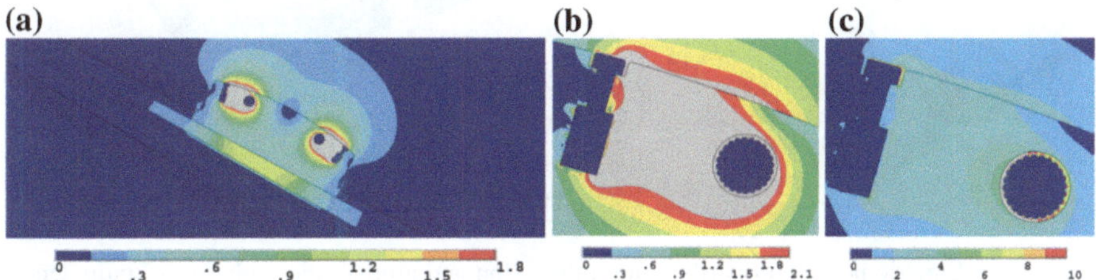

Fig. 4.9 Electric field contour plot around upper clamp, **a** electric field in gray area > 1.8 kV_{RMS}/mm, **b** electric field in gray area > (higher than the breakdown strength of standard air = 2.1 kV_{RMS}/mm), **c** around phase conductor, electric field in gray area > 10 kV_{RMS}/mm [21]

comply with electric field criteria on different parts of pylon. It was found that the heights of phase conductors from the unibody cross-arm should be higher than 2 m, which is not acceptable in terms of visual aspects of conductor clamps on the pylon. Therefore, steel enclosures with an appropriate diameter were utilized inside the non-conductive conductor clamps to reduce electric field intensities around and inside the clamps. The enclosures considerably decreased the height of conductor clamps. The potential distribution and equipotential surfaces around on the fully composite pylon with non-conductive conductor clamps and steel enclosure are displayed in Fig. 4.11. Finite element analysis in [21] revealed that electric field intensities at air inside the hollow cross-arm are higher than air breakdown strength. According to Fig. 4.12, electric field contour plot at a cutting plan through the upper conductor clamp proved that electric field intensity at the air around the internal ground cable is 3.08 kV_{RMS}/mm, which is higher than the breakdown strength of standard air (2.1 kV_{RMS}/mm).

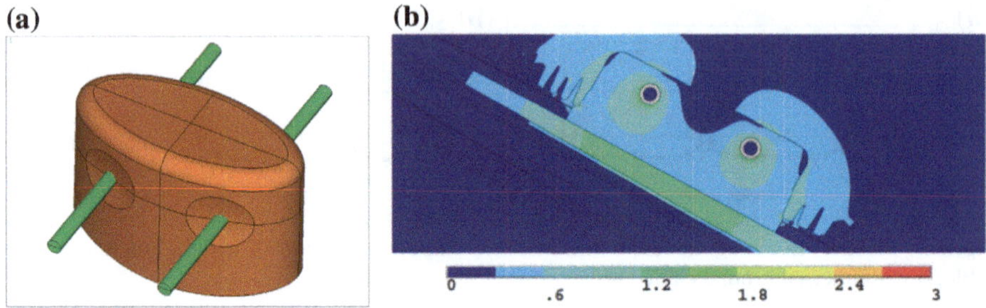

Fig. 4.10 **a** Simplified conductor clamp, **b** Electric field contour plot around clamp [21]

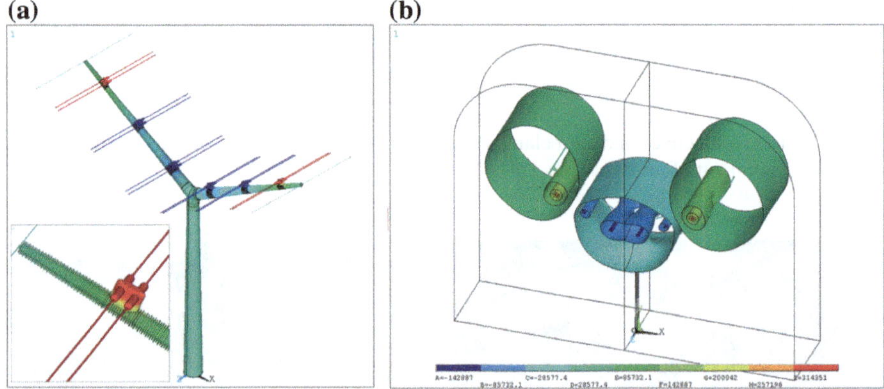

Fig. 4.11 **a** Electric potential distribution on the pylon and around upper phase, **b** Equipotential surfaces around the pylon [21]

Therefore, in the case of utilizing internal ground cable inside the unibody cross-arm, the air enclosed inside the cross-arm would be the source of local partial discharges and corona activities, which would lead to the degradation of cross-arm and thus, can cause the occurrence of puncture and flashover between ground cable and conductor enclosures under the energized parts [21]. As a result, it is concluded in [21] that the idea of using internal ground cable inside the unibody cross-arm of fully composite pylon should be discarded and an external ground connection should be considered for the pylon to provide ground potential access for two shield wires located at both tips of pylon. Finite element analyses in [21] revealed that the basic design of fully composite pylon has one critical weak point which is the exceedance of electric field magnitudes at air inside the unibody cross-arm from the air breakdown strength. This is due to the utilization of ground cable inside the unibody cross-arm. Therefore, some modifications should be made in the pylon design, which are discussed in the next section.

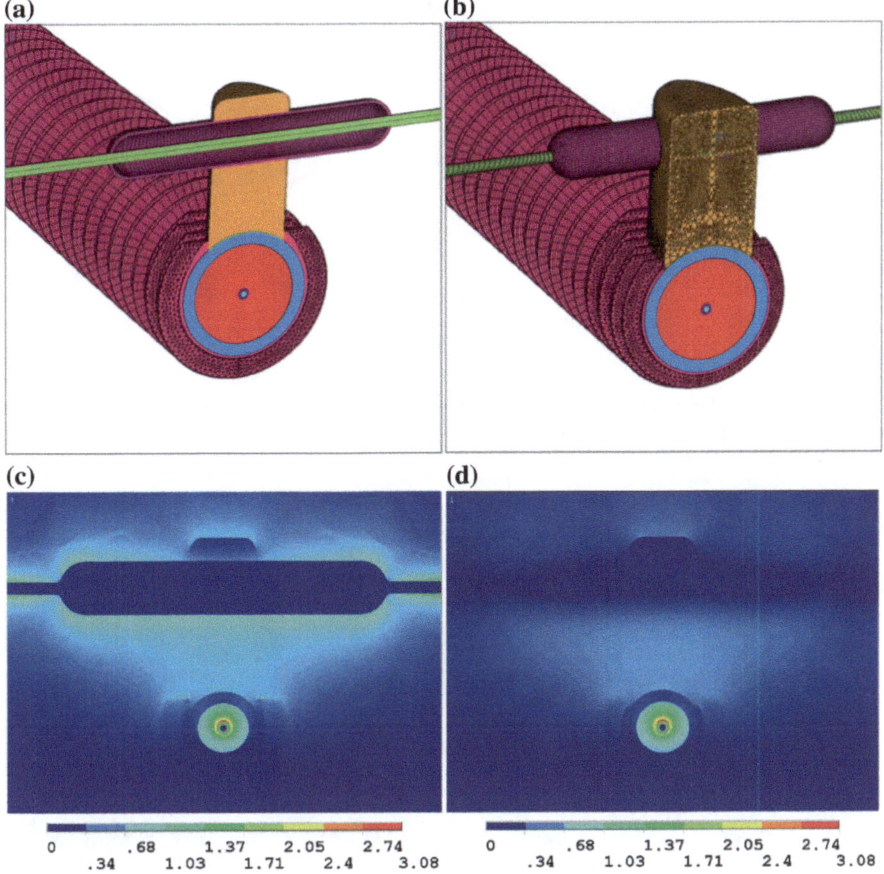

Fig. 4.12 **a** and **b** Cutting planes through a clamp for better visualization of air (red area), **c** and **d** electric field distribution on two cutting planes through the upper clamp [21]

4.4.2 Modifications in Fully Composite Pylon Design

The basic design of fully composite pylon utilized an internal ground cable through the hollow unibody cross-arm to connect shield wires to the earthing system as shown in Fig. 4.13a. It caused that air enclosed inside the unibody cross-arm to be exposed very high electric field intensities that was due to the short distances between energized parts (phase conductor) and grounded part (ground cable). Therefore, electric field studies in [21] proved that the connection between shield wires and earthing system must be provided externally, by using a ground connector illustrated in Fig. 4.13b, per example.

Since the ground connector has a ground potential, the direct distance between ground connector and phase conductors should be equal or higher than the minimum

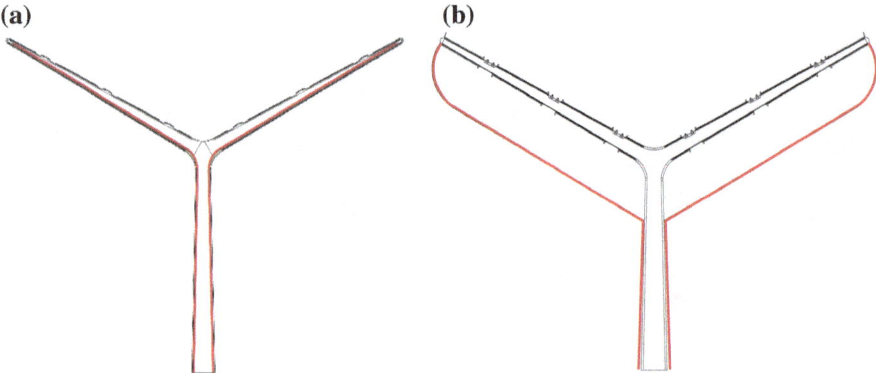

Fig. 4.13 **a** Internal ground cables (red colors) through the basic design of pylon, **b** external ground connectors (red colors) of modified pylon

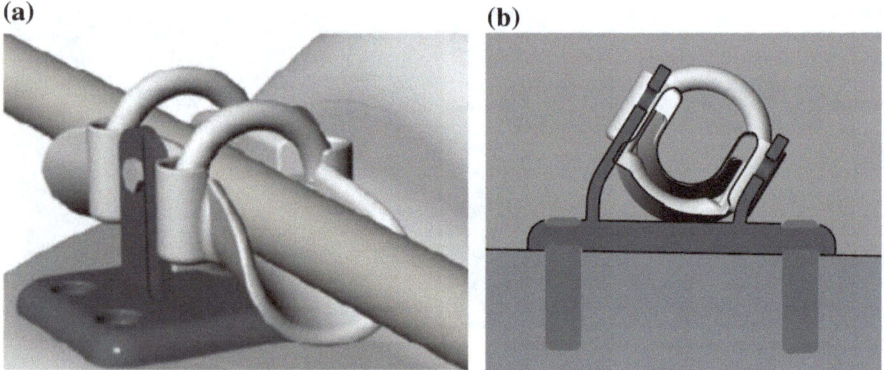

Fig. 4.14 **a** Three-dimensional view, **b** cross-sectional view of conductor clamp

air clearance between live part (phase conductor) and grounded part (shield wire). This distance is determined to be at least 2.80 m. Another benefit of utilizing the external ground connector is that the cross-section of unibody cross-arm is no longer exposed to high electric field stresses caused by phase-to-ground voltages. By taking away the ground potential from the unibody cross-arm, the electric field distribution around the attachment points on the cross-arm is completely modified. This helps the use of other alternatives for the conductor clamp design, which can be more feasible from a mechanical point of view.

Therefore, other modifications have been made in the basic design of pylon. Firstly, the shape of conductor clamp is changed to be as Fig. 4.14 and its material is decided to be metal (e.g. steel). By the use of full metal conductor clamps, material degradation around phase conductors due to higher electric field magnitudes no longer exists. Figure 4.14 shows that the new conductor clamp utilizes four metal bolts, which should be inserted into the cross-arm layer.

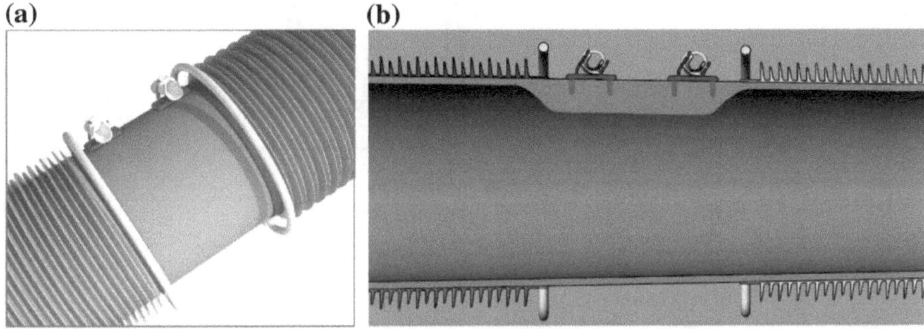

Fig. 4.15 **a** Three dimensional and **b** Cross-sectional view of corona rings near conductor clamps

Two important issues may arise from this modification. One is the presence of possible intensive electric field magnitudes at triple junctions where metal conductor clamp, air and cross-arm tube meet. Another is the electric field magnitudes around the bolts inside the cross-arm tube. The electric field magnitudes at these regions should be controlled to prevent internal partial discharges around the bolts and external partial discharges on conductor clamp near the triple junctions.

The solution for preventing internal partial discharges is utilizing a nut (for the bolt), which has a suitable diameter, and its end is rounded off. A cylindrical nut with flat end surfaces has sharp edges that can cause higher electric field intensities on the edges, therefore, the end surface's edges should be appropriately filleted to modify electric field magnitudes. Moreover, the space around the nut inside the cross-arm tube should be appropriately sealed by adhesive. Another modification in the basic pylon design is installing corona rings at both sides of conductor clamps, which is shown in Fig. 4.15.

Generally, corona rings are widely used in overhead lines' insulators at the voltage level of 220 kV and above to reduce electric field intensities near the triple junctions [18]. In the case of fully composite pylon, it is expected that utilizing two parallel corona rings with the same potential of phase (i.e. the potential of conductor clamps) drastically reduce electric field intensities on the conductor clamps, which are placed between the corona rings. From the electric field studies' experiences in [1, 21], the aims for using corona rings are:

- Reducing electric field intensities on metal conductor clamps especially near the triple junctions,
- Decreasing electric field intensities on sheds and sheath adjacent to the energized parts,
- Reducing electric field magnitudes inside the cross-arm layer (fiberglass tube) especially around the nuts of bolts,
- Decreasing electric field intensities at the air inside the cross-arm under the energized parts.

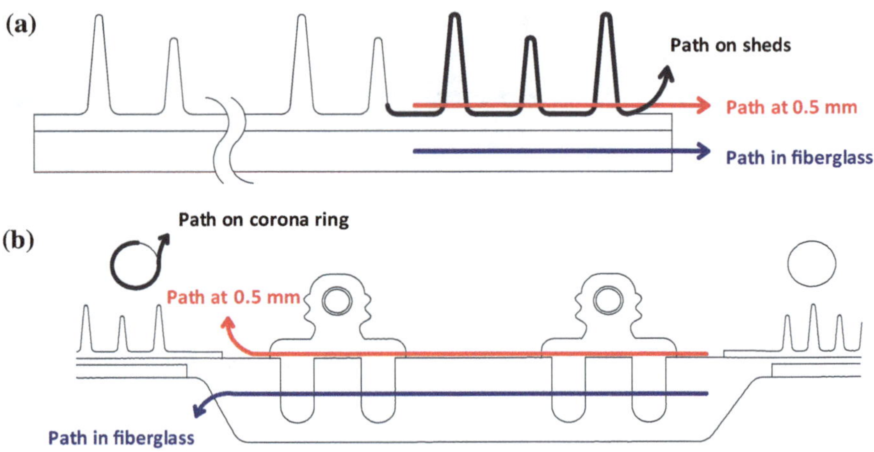

Fig. 4.16 Definition of paths **a** along insulation, **b** along conductor clamps' area and on corona rings

In this regard, a question is that what should be the dimensions and positions of corona rings, Because the positions and dimensions of corona rings may have a significant influence on electric field distribution around and inside the unibody cross-arm and conductor clamps (especially in triple junctions). Therefore, an optimum position and dimension for corona rings should be determined in a way that all electric field criteria on different parts of pylon should be satisfied. In the optimum case, electric field magnitudes on sheds, at 0.5 mm above shed housing, within fiber glass material, and at triple junctions are below the threshold values and the degradation of dielectric materials due to corona activities is unlikely. In the continuity of this chapter, the electric field performance of modified fully composite pylon is investigated by using finite element method.

4.4.3 Optimization of Corona Rings

Using ANSYS software, an optimization process has been carried out to find a design point in which local maximum electric field magnitudes on different parts of modified full composite pylon comply with all electric field criteria. In this regard, local maximum electric field magnitudes have been calculated in several regions of interest on the pylon. The regions of interest are as follows:

1. External paths along creepage distances on sheds, in the zones of 1, 2, 3, 4, 5 (Figs. 4.16 and 4.17).
2. Internal paths along the cross-arm inside fiberglass, at the zone of 1, 2, 3, 4, 5 (Figs. 4.16 and 4.17) and within the fiberglass layer around the nuts of bolts.

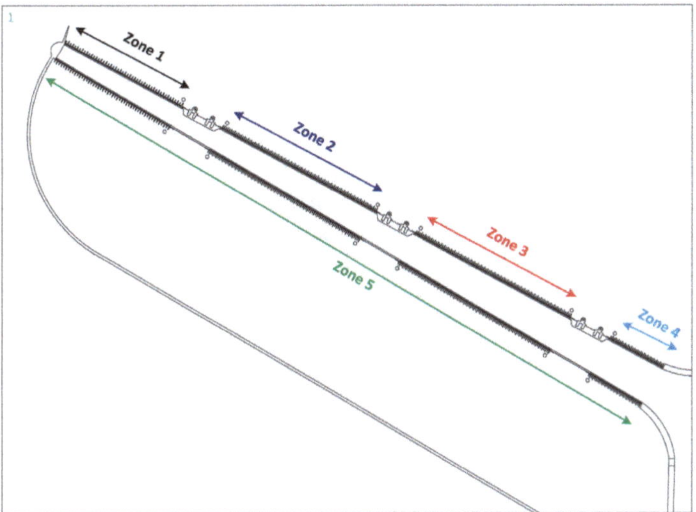

Fig. 4.17 Defined zones on the fully composite pylon

3. External paths at 0.5 mm above sheath surface, at the zones of 1, 2, 3, 4, 5—at the zones of upper, middle and lower conductor clamps (Figs. 4.16 and 4.17)
4. Paths on the circumference of corona rings, on the rings of R1, R1 sym, R2, R2 sym, S1, S1 sym, S2, S2 sym, T1, T1 sym, T2, T2 sym (Fig. 4.18)
5. Triple junction points around the conductor clamps, at the outer triple points 1, 4, 5, 8, 9 and 12, and—at the inner triple points of 2, 3, 6, 7, 10, 11, and at two triple points of 00 and 0 (Fig. 4.18)

First, some preliminary values are assigned to the horizontal position, vertical position and the diameter of corona rings, which are shown in Fig. 4.19a. The variations in design parameters are according to a three-level "do loop". The names and variations of parameters are as follows:

- Variation in the horizontal position of corona ring (ΔH): 0, 50, 100 mm
- Variation in the vertical position of corona ring (ΔV): 0, 50, 100 mm
- Variation in the diameter of corona ring (ΔD): 0, 15, 30 mm

Three parameters with three steps (variations) lead to 27 steps. It means that 27 finite element analyses have been done to find the best design point. The 27 alternatives for the corona ring are depicted in Fig. 4.19b. Figure 4.20d shows the variation of parameters with respect to the design point number. For example, design point number of 1 implies to the preliminary values of horizontal position, vertical position and the diameter of corona rings which have the variations' values of $\Delta H = 0$, $\Delta V = 0$ and $\Delta D = 0$. Design point number of 15 means that the horizontal position of corona ring is 200 mm + ΔH of 100 mm. With the same design point number, the vertical position of corona ring is 100 mm + ΔV of 50 mm and the

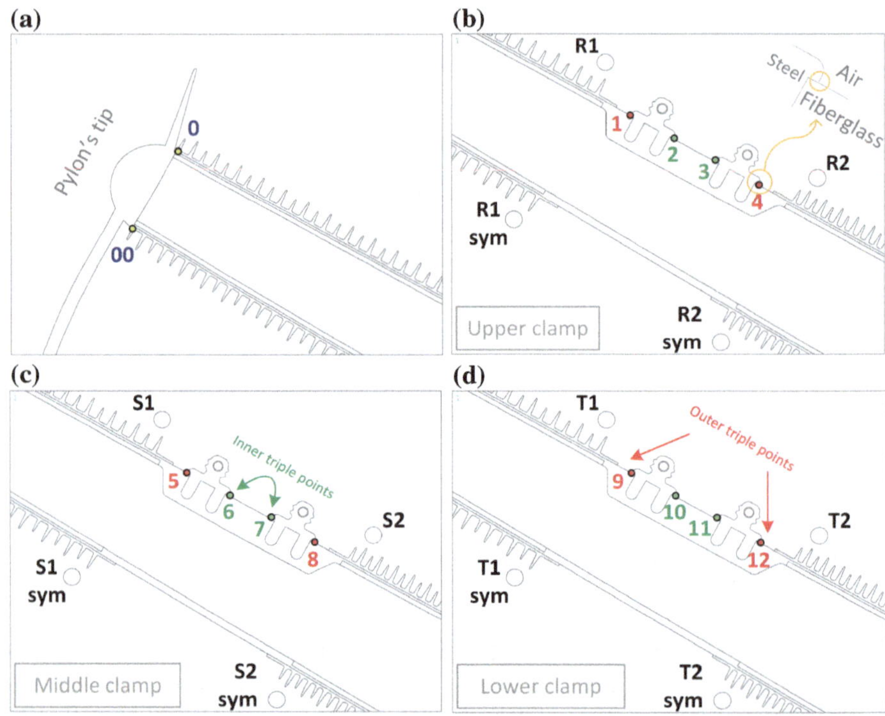

Fig. 4.18 Numbering of triple junction points and corona rings, **a** at pylon's tip, **b** around upper clamp, **c** around middle clamp, **d** around lower clamp

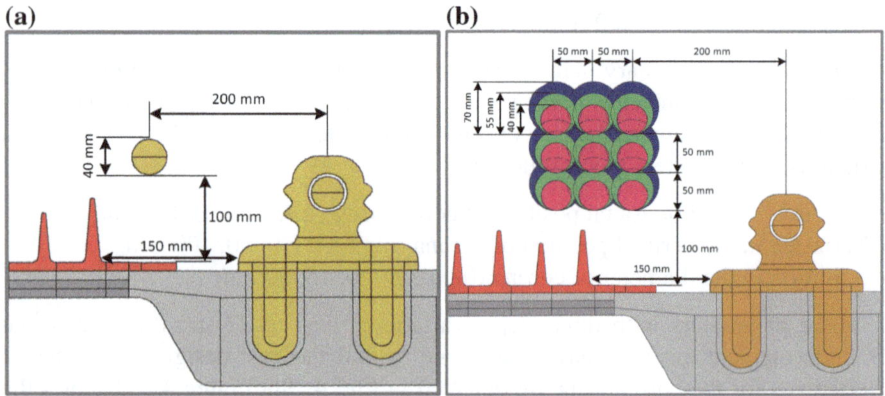

Fig. 4.19 **a** Preliminary position and diameter of corona ring, **b** 27 alternatives for the position and diameter of corona ring with three values for the each parameters of vertical position of corona ring, horizontal position of corona ring and diameter of corona ring

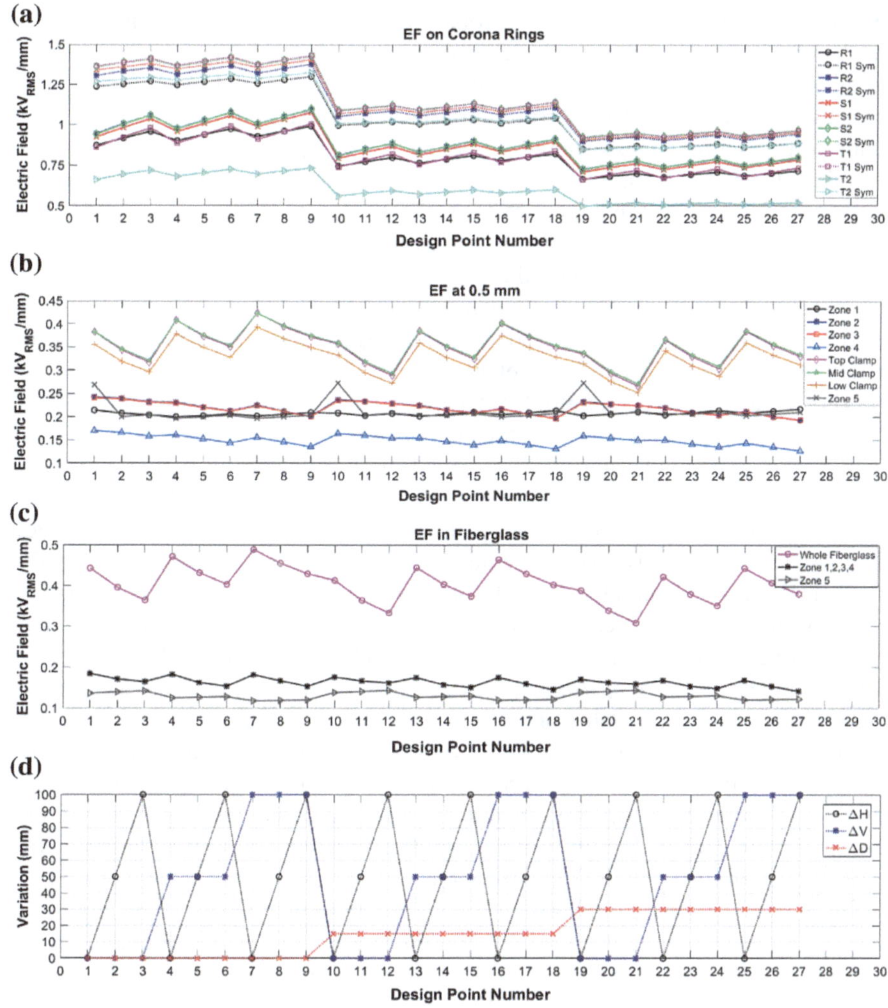

Fig. 4.20 Maximum electric field magnitudes on different regions of interest, **a** on corona rings, **b** at 0.5 mm above sheath surface, **c** within fiberglass material, **d** versus design point numbers

diameter of corona ring is 40 mm + ΔD of 15 mm. By this way, local maximum electric field magnitudes at different regions of interest are calculated and presented.

4.4.3.1 Maximum Electric Field Magnitudes on Different Regions of Interest

Maximum electric field magnitudes on different corona rings are calculated and shown in Fig. 4.20a for 27 design points. Figure 4.20a shows that the variations in horizontal and vertical positions of corona rings have not a significant influence on the electric field magnitudes on the corona rings. As it can be expected, increasing the diameter of corona rings significantly reduces electric field magnitudes on the corona rings. According to electric field criterion 2, maximum electric field magnitudes on the corona rings should be below 1.8 kV_{RMS}/mm. Figure 4.20a shows that this criterion has been fully satisfied with all design points because electric field magnitudes in all 27 design points are below 1.5 kV_{RMS}/mm. Maximum electric field magnitudes at 0.5 mm above sheath surface along five zones of interest and at 0.5 mm above the cross-arm layer around the conductor clamps are depicted in Fig. 4.20b. Figure 4.20b shows that there is no meaningful relation between the maximum electric field magnitudes at 0.5 mm above sheath surface along the five zones and the variations in the positions and diameters of corona rings.

However, electric field magnitudes in all regions of interest at 0.5 mm above sheath surface and cross-arm layer around the clamps are less than 0.45 kV_{RMS}/mm which means that electric field criterion 1 has been fully fulfilled for all design points.

Maximum electric field magnitudes within the cross-arm tube (fiberglass material) are shown in Fig. 4.20c for two paths along the cross-arm and for the whole fiberglass layer especially around the nuts of bolts. Electric field magnitudes within the fiberglass layer around the nuts of bolts significantly change with the variations in the positions and diameter of corona rings, but the variations in electric field magnitudes along two paths are negligible. Electric field magnitudes in these regions of interest are below 0.5 kV_{RMS}/mm, which is much lower than the electric field criterion 3 with the constraint value of 3 kV_{RMS}/mm. Therefore, electric field criterion 3 is satisfied within the fiberglass material for all corona ring design points. It should be mentioned that the limit of criterion 3 is related to a fiberglass tube (cross-arm) without any internal voids. It means that the fiberglass tube during production process should not contain any internal defects. Practically, this accuracy in production technology seems to be unlikely for the large structure of fully composite pylon [21]. Hence, electric field magnitudes within the fiberglass material should be kept much lower than the constraint value to avoid internal partial discharges. Reaching maximum electric field magnitudes below 0.5 kV_{RMS}/mm in fiberglass material implies that the electric field intensities inside fiberglass material are considerably lower than the constraint value of 3 kV_{RMS}/mm, thus this part of the design seems to be more feasible and practical. And due to much lower electric field intensities inside the fiberglass material, the possibility of internal partial discharges within the fiberglass material can be drastically reduced.

Local maximum electric field magnitudes on weather sheds along the creepage distances of five zones of interest are illustrated in Fig. 4.21a. Increasing vertical position of corona rings decreases electric field magnitudes on weather sheds. Variations in the diameter of corona rings have approximately no influence on the electric field magnitudes on weather sheds. In overall, it can be said that electric field intensities on weather sheds on the downside of cross-arm are higher than the upper side. However, it is mentioned in the explanation of electric field criteria that there are no direct electric field criteria on weather sheds. An electric field threshold value of 0.8 kV_{RMS}/mm is chosen for the regions of interest on weather sheds. According to

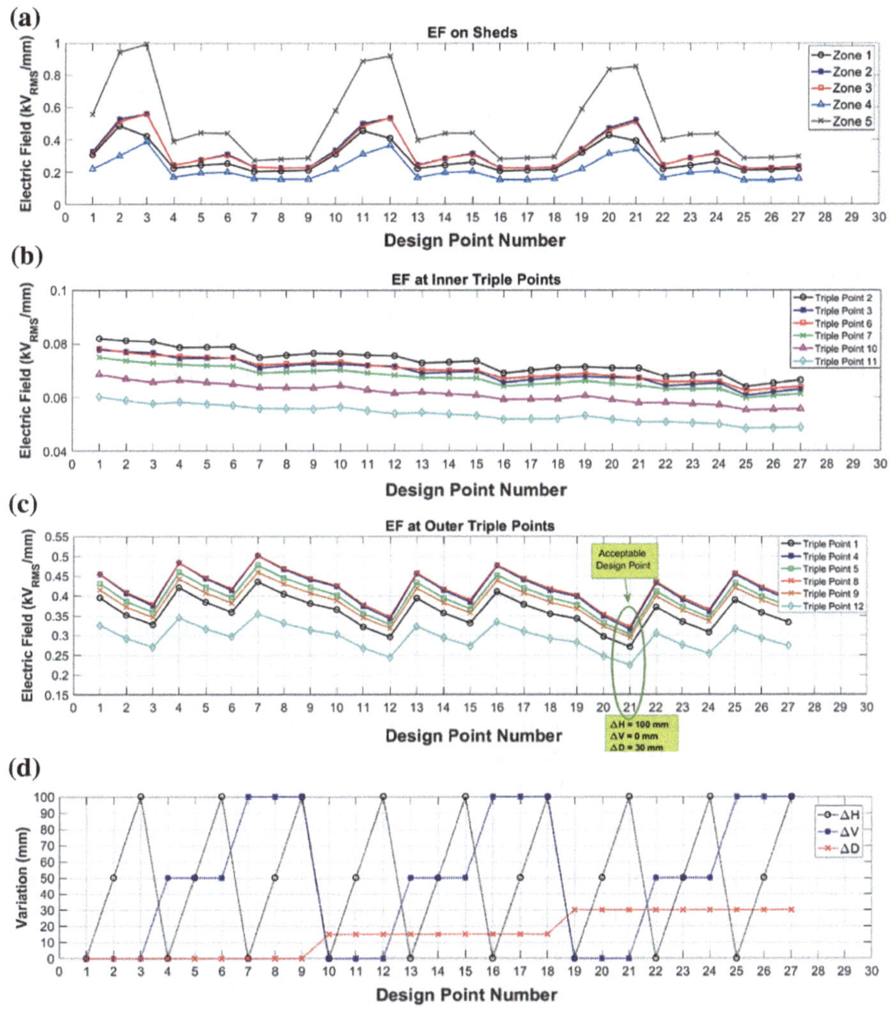

Fig. 4.21 Maximum electric field magnitudes on different regions of interest, **a** on weather sheds, **b** at inner triple points, **c** outer triple points, **d** versus design point number

Fig. 4.21a, all design points provide electric field magnitudes below 0.8 kV$_{RMS}$/mm except the design points with the numbers of 2, 3, 11, 12, 20 and 21.

Maximum electric field magnitudes at the inner triple points are shown in Fig. 4.21b. It can be observed that there is no specific relation between the design points (positions and diameter of corona rings) and electric field magnitudes at the above-mentioned triple junctions. The electric field magnitudes at these triple junctions are lower than 0.1 kV$_{RMS}$/mm. It means that electric field criterion 4 with the constraint value of 0.35 kV$_{RMS}$/mm has been satisfied in these inner triple points for all design points. On the other hand, electric field magnitudes at the outer triple points are shown in Fig. 4.21c. Figure 4.21c shows that the variations in the positions and diameter of corona rings impose different electric field magnitudes on these triple junctions. However, according to Fig. 4.21c, only design point number 12 and 21 provides electric field magnitudes below the constraint value of 0.35 kV_{RMS}/mm (criterion 4) for all the triple junctions.

The electric field magnitudes at some of the outer triple points in the design point of 12 are at the border of 0.35 kV_{RMS}/mm and for this reason, this design point is not appropriate and neglected. Therefore, only design point of 21 is chosen as an acceptable solution to keep electric field magnitudes below 0.35 kV_{RMS}/mm. And other design points are not suitable for the controlling of electric field intensities at triple junctions. As a result, only design point number 21 produces electric field magnitudes that satisfy the electric field criteria on different regions of interest including:

- On corona rings
- At 0.5 mm above sheath surface
- Within fiberglass material
- At triple junctions
- On weather sheds (except zone 5)

The zone 5 on the downside weather sheds has maximum electric field magnitude of 0.85 kV_{RMS}/mm. Electric field magnitude in this region is slightly higher than the adopted threshold value of 0.8 kV_{RMS}/mm. As it will be shown in the next section, electric field magnitudes on approximately all of weather sheds are much lower than the threshold value of 0.8 kV_{RMS}/mm except at the tips of some sheds, which have electric field magnitudes between 0.8 kV_{RMS}/mm and 0.85 kV_{RMS}/mm.

Nonetheless, controlling and reducing maximum electric field magnitudes at the triple junctions have a great importance, which has been fulfilled in the design process. Since electric field criteria in other regions of interest are also satisfied for the design point of 21, therefore, the design point of 21 can be introduced as the optimum design point for the modified fully composite pylon. The optimum design point corresponds to the corona rings with the horizontal position of 300 mm, the vertical position of 100 mm and the diameter of 70 mm, which are displayed in Fig. 4.22. These parameters' values are obtained from the optimization in FEM, however, electrical tests should be conducted on a prototype to verify the results of FEM.

4.4.3.2 Potential Distribution on the Fully Composite Pylon with Optimized Corona Rings

Finite element analyses show that it is feasible to design corona rings, which by using them, higher electric field magnitudes can be eliminated at different components of fully composite pylon especially at triple junctions. In this section, the electric field performance of fully composite pylon with optimized corona rings is presented in more details. The selected positions and diameter for the corona rings are considered in the pylon's design and an extensive finite element analysis was done to demonstrate

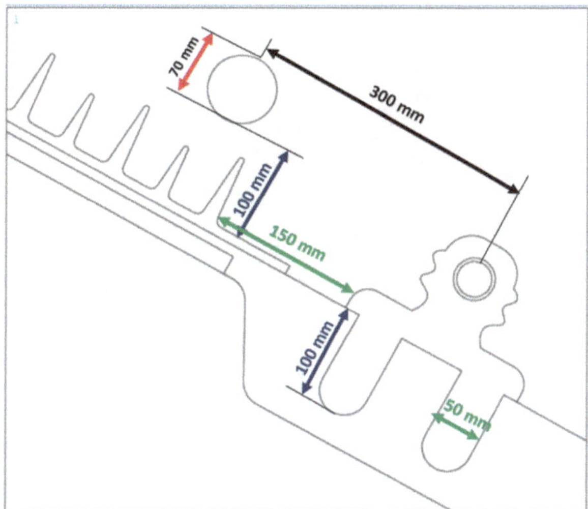

Fig. 4.22 Position and diameter of optimized corona rings

the potential and electric field distribution in the different parts of fully composite pylon.

Potential distribution around upper phase of fully composite pylon is shown in Fig. 4.23 for two different contour plots, one contour plot with 9 color bar and another contour plot with 128 color bar. It can be seen that the equipotential region (red area) around conductor clamps and corona rings in Fig. 4.23a is more extensive in comparison to Fig. 4.23b. Figure 4.23a shows that the red area at voltage level between 342.929 kV (voltage of phase A) and 285.774 kV covers the whole of the region surrounding the corona rings. The other equipotential regions are distinguished by orange and yellow colors. Figure 4.23b shows a more realistic view of potential distribution around the region. In Fig. 4.23b, the red area is concentrated on conductor clamps and corona rings, as well as the space between two conductor clamps. Moreover, the region between corona rings contains different colors, which means that the electric potential values in the region are lower than the applied voltage (342.929 kV) to the conductor clamps and corona rings. However, as it will be shown, the contour plots with the lower number of color bars are used to detect the regions with high electric field intensities.

Potential distribution on the top of fully composite pylon is illustrated in Fig. 4.24. As the contour plot (with 128 color bar) shows in Fig. 4.24, upper phases on both sides of pylon have the peak voltage of 342.929 kV. Other phases have the half of the peak voltage, −171.464 kV, with ±120° difference in phase angles. The blue colors represent the voltage of −171.464 kV which has been loaded to the middle and lower phases. The transition from the peak voltage of 342.929 kV (on upper phases) to the ground potential (on shield wires and external ground connector) are represented by changing color from red to yellow, green and cyan. The same approach can be imagined for the potential distribution between upper phase and middle phase. At

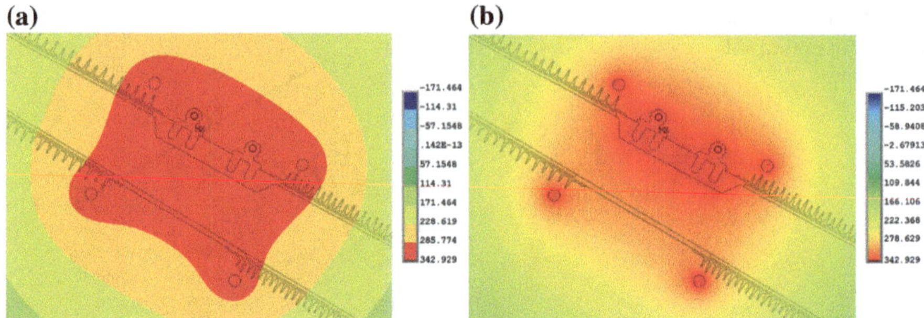

Fig. 4.23 Contour plot of potential distribution around upper phase with **a** 9 color bar, **b** 128 color bar

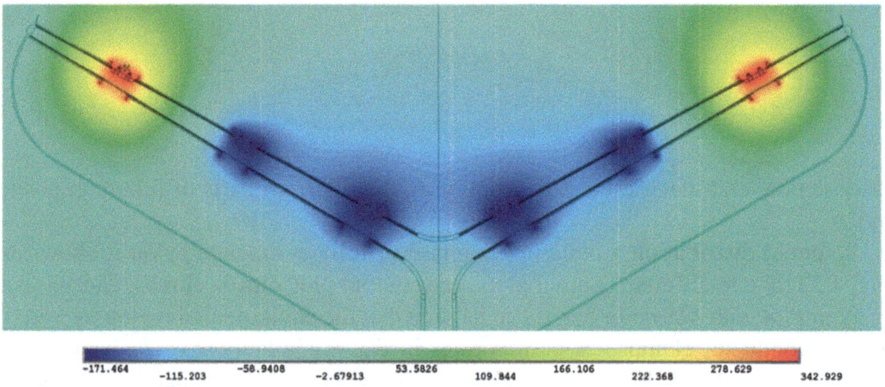

Fig. 4.24 Potential distribution on the top of fully composite pylon

the instantaneous time that middle and lower phases have the same electric potential (−171.464 kV), the color between these phases has been affected by the ground connector with ground potential. The color between two lower phases at the center of the pylon also varies in the same way.

In order to better represent the effect of ground connector on the potential distribution, the equipotential lines around fully composite pylon are displayed in Fig. 4.25. Figure 4.25 shows that equipotential lines caused by energized clamps and corona rings are restricted by the external ground connector. This restriction is only valid for this cross-section of fully composite pylon. Figure 4.25 reveals the higher density of lines between the upper phase and middle phase whereas there is a lower number of equipotential lines between middle phase and lower phase, as well as between two lower phases, which cannot be seen directly in the contour plots of potential distributions. Figure 4.25 also shows that utilizing a metal flange with floating potential (at the center of pylon) can distort the equipotential lines between the flange and ground connectors at the center of pylon. It can also be seen that equipotential lines at the top of cross-arm can easily distribute in the free space above the cross-arm.

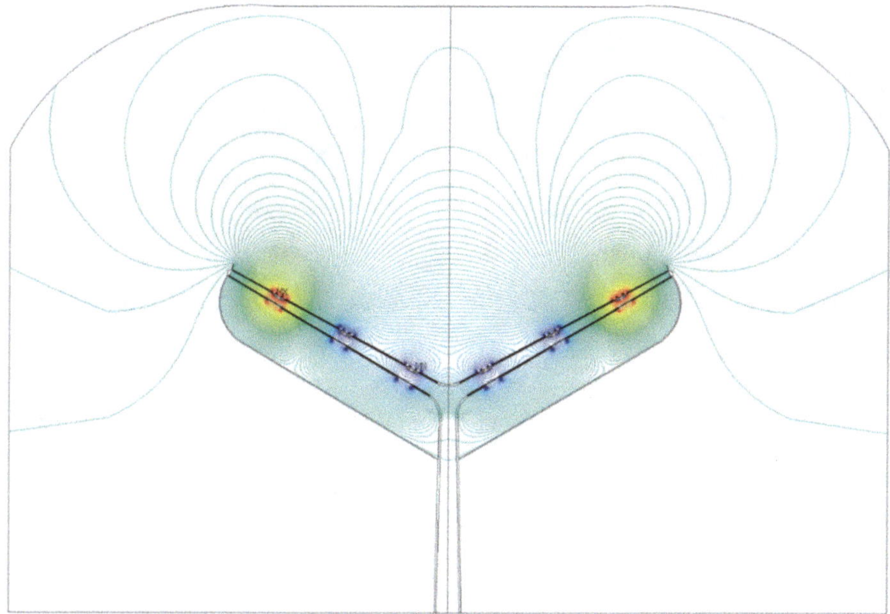

Fig. 4.25 Equipotential lines around fully composite pylon

Some equipotential lines in both sides of pylon have a trend to be closer to the earth surface, which is due to the ground potential of earth surface.

4.4.3.3 Electric Field Distribution on the Fully Composite Pylon with Optimized Corona Rings

Electric field distributions around the upper phase are shown in Fig. 4.26 for two different contour plots with 9 and 128 color bars. The upper phase is selected because it is exposed to phase-to-phase voltage stress (between upper phase and middle phase) as well as phase-to-ground voltage stress (between upper phase and shield wire and external ground connector). Other phases are also in the exposure of phase-to-ground voltage stresses due to proximity with the external ground connector. However, electric field contour plot in Fig. 4.26a shows a blue region between the corona rings in which electric field magnitudes are between 0 and 0.106 kV_{RMS}/mm. This gives an insight vision in the design of fully composite pylon. It means that utilizing corona rings around conductor clamps effectively reduces electric field intensities within the region enclosed by the corona rings. Figure 4.26a with lower color bar explicitly indicates that higher electric field intensities are removed from the surfaces of conductor clamps and transmitted on the corona rings. Therefore, the corona rings in the pylon design are an inseparable part of the pylon. Lower cross-sections of corona rings in Fig. 4.26a sustain higher electric field intensities (red

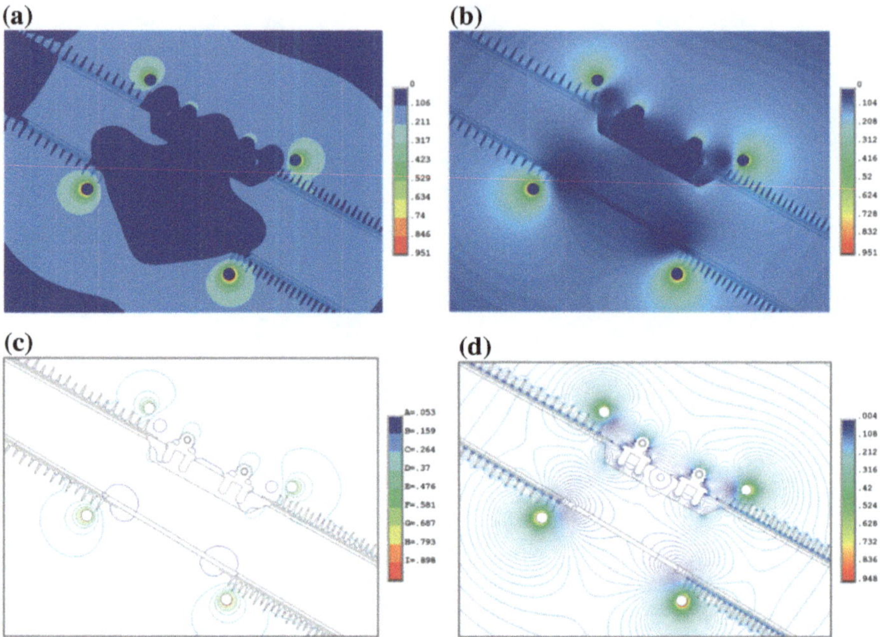

Fig. 4.26 Electric field cloud chart around upper phase with **a** 9 color bar and **b** 128 color bar, and electric field distribution lines with **c** 9 color bar and **d** 128 color bar

color on them) with respect to the upper cross-section of the corona rings. It is due to the shorter distance between the lower cross-section of corona rings and the external ground connector. As explained before, this distance is equal to the phase-to-ground air clearance on the fully composite pylon. However, Fig. 4.26b gives less detail regarding the electric field distribution around the upper phase. The higher number of color bars provides a complicated condition to evaluate and detect the regions with enhanced electric field magnitudes.

From Fig. 4.26a and b, it can hardly be seen that electric field magnitudes on two outer triple junctions are in the range of 0.106–0.211 kV_{RMS}/mm, which are below the constraint value of 0.35 kV_{RMS}/mm. Electric field distribution lines with different color bars are shown in Fig. 4.26c and d to give more detail for the electric field distribution within the region. For example, it can easily be observed in Fig. 4.26d that there are some electric field lines around the nuts of conductor clamps, which are placed inside the fiberglass layer. However, the magnitudes of these lines are much lower than the constraint value of 3 kV_{RMS}/mm.

Actually, Fig. 4.27 shows that the highest electric field intensity on different regions of fully composite pylon is less than 0.951 kV_{RMS}/mm which is specified by the value of red color in color bar. The higher electric field magnitudes are concentrated on the corona rings and are more visible in Fig. 4.27. The electric field intensities at the air around the fully composite pylon are also lower than the highest

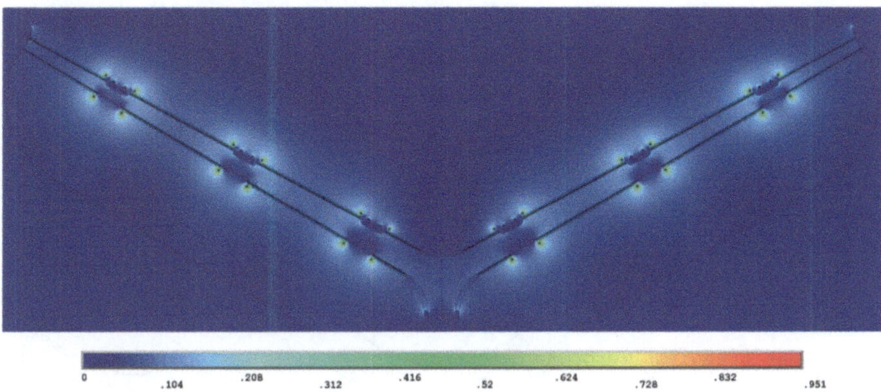

Fig. 4.27 Electric field cloud chart on the top of fully composite pylon

electric field magnitude of 0.951 kV_{RMS}/mm. By considering the value of 2.1 kV_{RMS}/mm for the breakdown strength of standard air, the air around fully composite pylon will not face with the air insulation breakdown. It can also be seen that the interaction between energized parts (especially corona rings) with the grounded parts (external ground connectors and shield wires) has been demonstrated by the transition from light colors to blue color (with the electric field magnitude of zero). The contour plots of electric field give an insight knowledge about the overall range of electric field intensities on the areas of interest. In order to get more information regarding the magnitudes of electric field on different components such as on corona rings, weather sheds, at 0.5 mm above sheath surface and within fiberglass material, several paths have been defined to derive the variation of electric field along the paths. In order to get electric field magnitudes on the corona rings, two paths have been defined on the circumference of each corona ring; one path on the upper cross-section of corona ring (e.g. R1) and another on the lower cross-section of corona ring (e.g. R1 Sym). Spatial distributions of electric field on the circumference of corona rings are shown in Fig. 4.28. According to Fig. 4.28, maximum electric field magnitudes on the corona rings are below 0.951 kV_{RMS}/mm. Spatial distributions of electric field on the upper cross-section of corona rings are in the form of sinusoidal, whereas spatial distribution waveforms on the lower cross-section of corona rings have been distorted due to the proximity of them with the external ground connector. It had been found from corona optimization process that the optimum design point provides acceptable performance on corona rings. The results of the optimum design point are presented here to demonstrate that maximum electric field magnitudes on all corona rings (i.e. 0.951 kV_{RMS}/mm) are below the constraint value of 1.8 kV_{RMS}/mm (electric field criterion 2).

Electric field magnitudes at 0.5 mm above sheath, is one of the design criteria, which should be fulfilled, are illustrated in Fig. 4.29. Figure 4.29a is related to the path lengths along the unibody cross-arm for the different zones specified in Fig. 4.17. It can be seen that the electric field magnitudes at zone 1, 2, 3, 4 and at

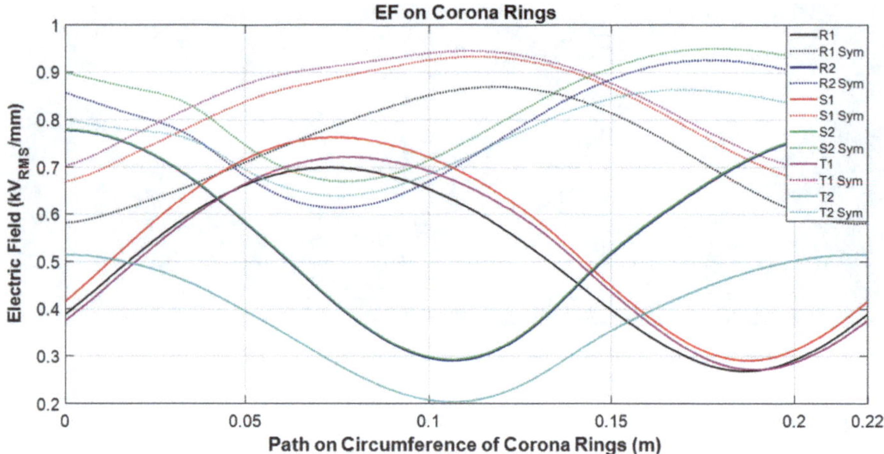

Fig. 4.28 Spatial distributions of electric field on the circumference of corona rings

the zones of top clamp, middle clamp and lower clamp are lower than the constraint value of 0.45 kV_{RMS}/mm. The regions with zero electric field are related to the metal conductor clamps, which have intersections with the path at 0.5 mm above sheath surface. Electric field magnitudes at 0.5 mm above sheath surface at zone 5 is shown in Fig. 4.29b. According to Fig. 4.29b, electric field magnitudes at zone 5 along the unibody cross-arm is also below 0.45 kV_{RMS}/mm. Therefore, as it is indicated in optimization process results, the electric field magnitudes at 0.5 mm above sheath surface of fully composite pylon satisfy the electric field criterion 1 (with the constraint value of 0.45 kV_{RMS}/mm).

Electric field magnitudes along two paths within fiberglass layer (cross-arm tube) are depicted in Fig. 4.30. The magnitudes of electric field in the zones of 1, 2, 3, 4 and 5 are much lower than the constraint value of 3 kV_{RMS}/mm specified in electric field criterion 3. It means that the presence of any void during manufacturing of unibody cross-arm tube may not cause internal partial discharge inside the fiberglass material because of the very low electric field magnitudes (below 0.16 kV_{RMS}/mm) within the fiberglass layer. However, Fig. 4.30 shows that electric field magnitudes have higher values around the corona rings and have lower values between the corona rings of each phase.

Electric field distributions along the creepage distances on the unibody cross-arm are depicted in Fig. 4.31 for different zones of 1, 2, 3, 4 and 5. Figure 4.31a shows that maximum electric field magnitudes on weather sheds at the zones of 1, 2, 3, 4 are below 0.6 kV_{RMS}/mm which are lower than the threshold value of 0.8 kV_{RMS}/mm. Electric field magnitudes on the tip of sheds are higher at the regions near the corona rings. Since the shed housing along the unibody cross-arm does not have sheds between two corona rings of each phase, therefore, the curves in Fig. 4.31a are discrete and there is no value in these regions. On the other hand, Fig. 4.31b shows the electric field magnitudes at the zone of 5. It can be seen that

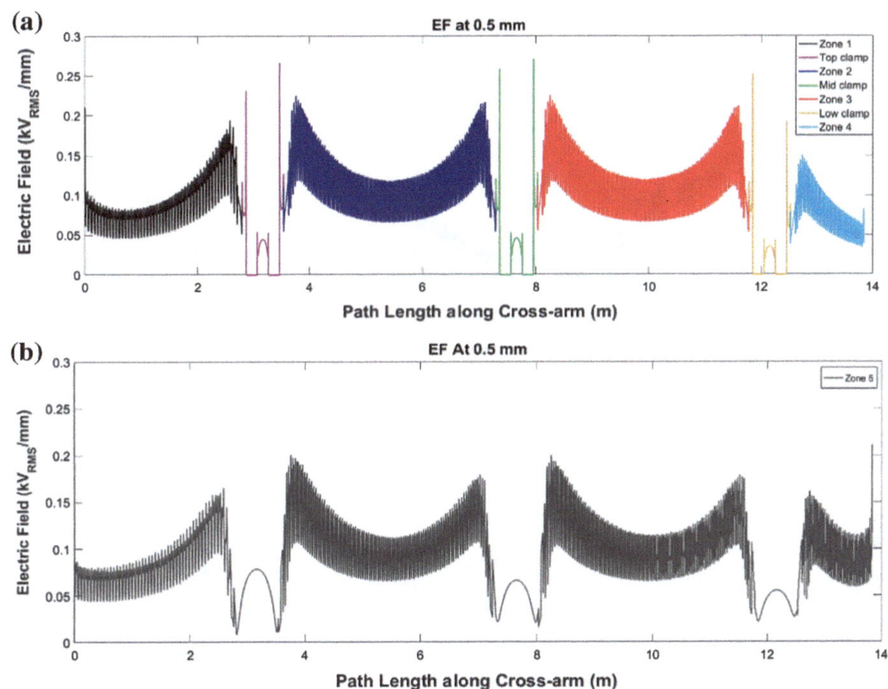

Fig. 4.29 Electric field magnitudes at 0.5 mm above sheath surface at, **a** zone 1, 2, 3, 4 and at the zones of Top clamp, Middle clamp and Lower clamp, **b** zone 5

some of the tips of weather sheds sustain higher electric field magnitudes around and under corona rings. In zone 5, all weather sheds are in the exposure of electric field magnitudes below 0.8 kV_{RMS}/mm except at the tip of one shed which has electric field magnitude between 0.8 and 0.85 kV_{RMS}/mm. It had been mentioned in the corona ring optimization process that the electric field intensities in one of the shed regions

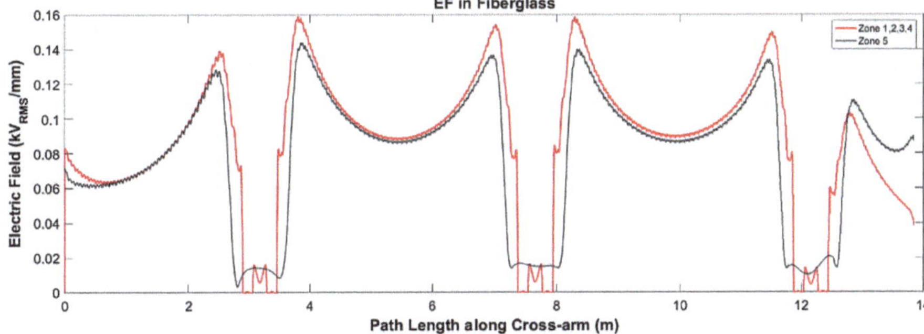

Fig. 4.30 Electric field magnitudes along two paths within fiberglass layer

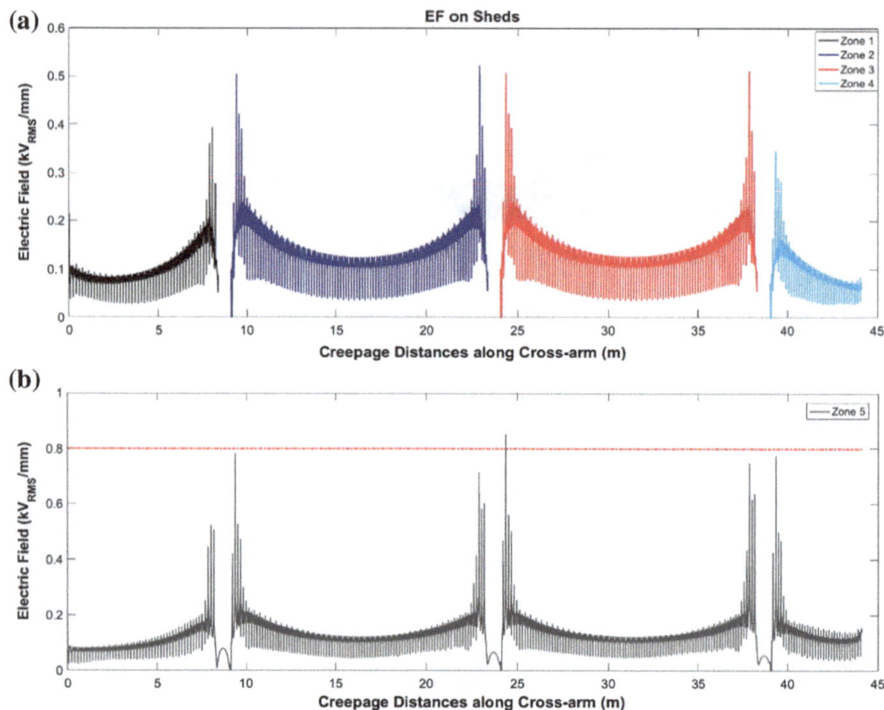

Fig. 4.31 Electric field distribution on sheds (along creepage distances) at the, **a** zone 1, 2, 3, 4, **b** zone 5

exceed the threshold value of 0.8 kV_{RMS}/mm. Here, it is revealed that only the tip of one weather shed is in the exposure of higher electric field magnitude. By trading off between the exceedance of electric field on triple junctions and only one weather shed, it is completely clear that the selection of optimum design point based on acceptable performance at triple junctions is more reasonable and have considerable importance in comparison to the exceeding of electric field on only one weather shed. Consequently, it can be concluded that utilizing selected optimum design point for the fully composite pylon provides excellent electric field performance for the pylon. Because all of the mentioned electric field criteria have been fulfilled in the electric field design process of fully composite pylon.

4.5 Summary

In this chapter, the required creepage distances between the upper phase and shield wire as well as phase-to-phase regions were calculated. The application of different shed profiles in insulators' shed housing designs were reviewed and, consequently, a small and large shed profile were allocated for the shed housing on the unibody cross-arm. Numerous finite element analyses of the pylon were carried out to evaluate electric field and potential distribution around and inside the unibody cross-arm. Two design scenarios of utilizing internal and external ground connection for the shield wires were examined. It was observed that utilization of internal ground cable inside the hollow unibody cross-arm is not feasible because the air inside the unibody cross-arm sustains higher electric field magnitudes than air insulation breakdown strength, which may lead to insulation failure. Therefore, an external ground connection was used for providing the ground potential for shield wires. Different conductor clamp designs with various material in used were considered for the attachment points between phase conductors and the unibody cross-arm. Non-conductive conductor clamps with steel bolts provide floating potential on the bolts, which cause air insulation breakdown at the air around the bolts. As a feasible solution, a steel conductor clamp was proposed, which provides better conditions for the connection and electric field considerations. Computation of electric field magnitudes in the different regions of interest on the unibody cross-arm showed that electric field magnitudes at triple junctions of steel conductor clamp are higher than the threshold values. In order to reduce the electric field magnitudes at these points, the idea of utilizing corona rings at both sides of conductor clamps were introduced. For this reason, an optimization process was done to find appropriate dimension and positions for the corona rings. In the optimum case, electric field magnitudes on different parts of the unibody cross-arm (on corona rings, at 0.5 mm above shed housing, on sheds, within fiberglass tube, at triple junctions and at the air inside the cross-arm) were below the threshold values and therefore, the outcome of the optimization was a design point in which fully composite pylon represents an acceptable electric field performance in all different regions of interests.

References

1. T. Jahangiri, C.L. Bak, F.M.F. da Silva, B. Endahl, Electric field and potential distribution in a 420 kV novel unibody composite cross-arm, in *Proceedings of the 24th Nordic Insulation Symposium on Materials, Components and Diagnostics* (2015), pp. 111–116
2. K.O. Papailiou, *CIGRE Green Book: Overhead Lines* (Springer, 2017)
3. IEC 815, Guide for the selection of insulators in respect of polluted conditions (1986)
4. IEC/TS 60815-1, Selection and dimensioning of high-voltage insulators intended for use in polluted conditions—Part 1: definitions, information and general principles (2008)

5. IEC/TS 60815-2, Selection and dimensioning of high-voltage insulators intended for use in polluted conditions—Part 2: ceramic and glass insulators for a.c. systems (2008)
6. IEC/TS 60815-3, Selection and dimensioning of high-voltage insulators intended for use in polluted conditions—Part 3: polymer insulators for a.c. systems (2008)
7. Z. Chu-yan, W. Li-ming, J. Zhi-dong, G. Zhi-cheng, Research on the optimization of UHV AC composite insulators' shed, in *2010 International Conference on High Voltage Engineering and Application (ICHVE)* (2010), pp. 345–348
8. F. Zhang, L. Wang, Z. Guan, M. Macalpine, Influence of composite insulator shed design on contamination flashover performance at high altitudes. IEEE Trans. Dielectr. Electr. Insul. **18**(3), 739–744 (2011)
9. Q. Hu, L. Shu, X. Jiang, C. Sun, Z. Zhang, J. Hu, Effects of shed configuration on AC flashover performance of ice-covered composite long-rod insulators. IEEE Trans. Dielectr. Electr. Insul. **19**(1), 200–208 (2012)
10. C. Zachariades, *Development of an Insulating Cross-arm for Overhead Lines* (The University of Manchester, 2014)
11. ABB Co., Hollow composite insulators 72-1,200 kV design for reliable performance
12. A.J. Phillips, D.J. Childs, H.M. Schneider, Aging of non-ceramic insulators due to corona from water drops. IEEE Trans. Power Deliv. **14**(3), 1081–1089 (1999)
13. C. Zachariades, S.M. Rowland, I. Cotton, V. Peesapati, D. Chambers, Development of electric-field stress control devices for a 132 kV insulating cross-arm using finite-element analysis. IEEE Trans. Power Deliv. **31**(5), 2105–2113 (2016)
14. W. Que, *Electric Field and Voltage Distributions Along Non-Ceramic Insulators* (The Ohio State University, 2002)
15. U. Schümann, F. Barcikowski, M. Schreiber, H.C. Kärner, J.M. Seifert, FEM calculation and measurement of the electrical field distribution of HV composite insulator arrangements, in *CIGRE Session* (Paris, 2002), pp. 33–404
16. A.J. Phillips et al., Electric fields on AC composite transmission line insulators. IEEE Trans. Power Deliv. **23**(2), 823–830 (2008)
17. R. Lings, *EPRI AC Transmission Line Reference Book—200 kV and Above*, 3rd edn. (2005)
18. R. Anbarasan, S. Usa, Electrical field computation of polymeric insulator using reduced dimension modeling. IEEE Trans. Dielectr. Electr. Insul. **22**(2), 739–746 (2015)
19. V. Peesapati et al., 3D electric field computation of a composite cross-arm, in *IEEE International Symposium on Electrical Insulation (ISEI)* (2012), pp. 464–468
20. X. Yang, N. Li, Z. Peng, J. Liao, Q. Wang, Potential distribution computation and structure optimization for composite cross-arms in 750 kV AC transmission line. IEEE Trans. Dielectr. Electr. Insul. **21**(4), 1660–1669 (2014)
21. T. Jahangiri, Q. Wang, C. L. Bak, F. M. F. da Silva, H. Skouboe, Electric stress computations for designing a novel unibody composite cross-arm using finite element method, in *IEEE Transactions on Dielectrics and Electrical Insulation*, vol. 24, no. 6 (2017)
22. J. Li, Z. Peng, X. Yang, Potential calculation and grading ring design for ceramic insulators in 1000 kV UHV substations. IEEE Trans. Dielectr. Electr. Insul. **19**(2), 723–732 (2012)
23. X. Yang, Q. Wang, H. Wang, S. Zhang, Z. Peng, Transient electric field computation for composite cross-arm in 750 kV AC transmission line under lightning impulse voltage. IEEE Trans. Dielectr. Electr. Insul. **23**(4), 1942–1950 (2016)
24. Q. Wang, X. Yang, H. Li, Z. Guo, Z. Peng, Computation and analysis of the electric field distribution and voltage-sharing characteristics for 1000 kV composite tower, in *IEEE International Conference on Dielectrics (ICD)* (2016)
25. D. Nie, H. Zhang, Z. Chen, X. Shen, Z. Du, Optimization design of grading ring and electrical field analysis of 800 kV UHVDC wall bushing. IEEE Trans. Dielectr. Electr. Insul. **20**(4), 1361–1368 (2013)

26. S. Zhang, Z. Peng, P. Liu, N. Li, Design and dielectric characteristics of the ±1100 kV UHVDC wall bushing in China. IEEE Trans. Dielectr. Electr. Insul. **22**(1), 409–419 (2015)
27. V. Peesapati et al., Electric field computation for a 400 kV composite cross-arm, in *Conference on Electrical Insulation and Dielectric Phenomena (CEIDP)* (2012)
28. BYSTRUP Power Pylon Design (Online), https://www.powerpylons.com/. Accessed 13 Apr 2018

Chapter 5
Electric Field Verification by High Voltage Experiments on the Composite Cross-Arm

5.1 Introduction

Corona discharge on the polymeric surface of a composite insulator/cross-arm is the main reason for the component's aging, since by-products and high temperature generated during corona discharge lead to loss of hydrophobicity, erosion and tracking on the polymeric surface [1]. Corona discharge in dry conditions might be prevented in advance by imposing restrictions on electric field distribution. However, corona discharge due to water drops is unpredictable. Thus, in the design stage of a composite insulator/cross-arm, both electric field calculation as well as experiments should be done for the verification of corona activities on its surface.

5.1.1 Fundamental of Corona Discharge

Corona discharge is a kind of external partial discharge (PD). It usually happens in gaseous media which is surrounding an electrode at high voltage potential. Essentially, corona discharge is an electron avalanche, initiated by local electric field concentration that is higher than the electrical strength of the surrounding gas which is ionized consequently.

According to voltage types at the electrode, corona discharge can be divided into DC corona and AC corona.

5.1.1.1 DC Corona

DC corona is categorized into positive corona and negative corona, which indicates the high voltage electrode are at positive and negative potential, respectively [2]. Figures 5.1 and 5.2 indicate a positive and negative DC corona discharge initiated

T. Jahangiri et al., *Electrical Design of a 400 kV Composite Tower*, Lecture Notes in Electrical Engineering 557,
https://doi.org/10.1007/978-3-030-17843-7_5

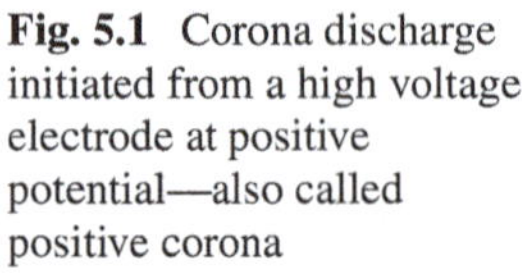

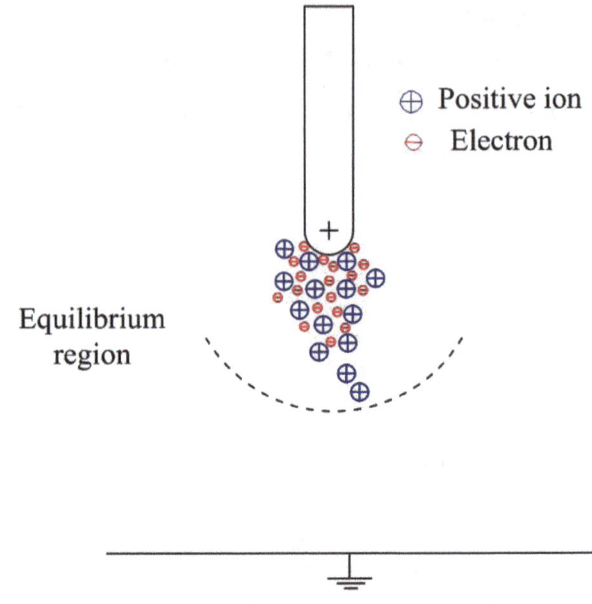

Fig. 5.1 Corona discharge initiated from a high voltage electrode at positive potential—also called positive corona

in a rod-plane electrode arrangement, respectively. According to the mechanism of electron avalanche, as long as initial electrons are generated by the local reinforced electric field, they are driven by the electric field to the positive polarity or the ground and generate more and more charge particles by colliding with neutral gas molecules continuously. Meanwhile, in theirs traveling paths, electrons either attach to electronegative gas molecules (such as O_2 in the air), forming negative ions, or recombine into neutral molecules with positive ions. Both process impede the development of the electron avalanche, i.e. corona discharge. There is an equilibrium region, also called boundary surface, between electrodes where ionization rate, represented by ionization coefficient α, equals with attachment rate, represented by attachment coefficient η, which indicates cease of the corona discharge. For a specific insulating gas, α and η depends on applied electric field E and the gaseous pressure p, which means a corona discharge is dependent on E and p.

In Fig. 5.1, the electron avalanche is initiated from the equilibrium region and heading to the electrode (anode). While in Fig. 5.2 it is initiated from the electrode (cathode) and heading to the ground.

According to different visual and electrical characteristics, positive corona discharge is further divided into four modes, including burst corona, onset streamer, positive glow and breakdown streamers, which happens in sequence with a increasing applied voltage [2]. Similarly, negative corona is divided into Trichel streamer, negative pulseless glow and negative streamer, respectively [2].

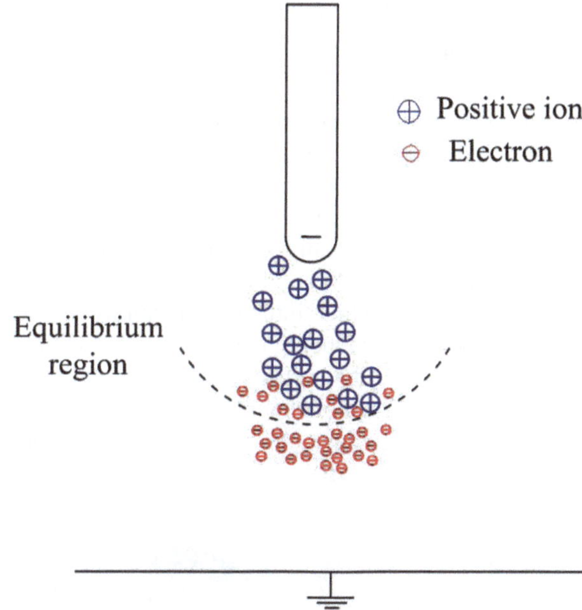

Fig. 5.2 Corona discharge initiated from a high voltage electrode at negative potential—also called negative corona

5.1.1.2 AC Corona

With a AC voltage whose polarity and magnitude varying, different corona modes are presented in a cycle of the applied voltage. Also, corona mode varies according to different gap lengths. For shorter gap, burst corona, positive onset streamer and Trichel streamers are present; While for longer gap, negative/positive glow discharges, positive breakdown streamers and Trichel streamers are observed [2].

Corona discharge in high voltage engineering field is inevitable, however it is undesired in most cases. For instance, corona discharge always happen on the surface of overhead lines where electric field is concentrated, which leads to power loss as well as radio interference. In addition, corona discharge generates high-frequency electromagnetic radiation which causes panic from residents nearby thus leads to rejection to new overhead lines.

Corona discharge happens on solid insulation if a local electric field intensifying, possibly caused by deposit of dust particles, rain drops and contamination, is present. In this case, especially when water is present, high temperature and by-products, such as ozone and acid, that generated during corona discharge cause insulation material damage.

Consequently, engineers take great efforts to optimize the electrical design of components in high voltage engineering to avoid local electric field concentration, thus avoid corona discharges.

5.1.2 Corona Discharge on the Surface of a Composite Insulator

Weather sheds of a composite insulator are utilized to provide enough creepage distance and protection for the internal Fiber Reinforced Plastic (FRP) rod. They cover a wide range of polymeric materials, including polydimethylsiloxane (SIR), ethylene propylene diene (EPDM), alloys of EPDM and silicone, ethylene vinyl acetate (EVA) and cycloaliphatic and aromatic epoxy resins [3]. Nowadays, rubbers based on SIR and EPDM are dominant in the weather sheds materials.

Hydrophobicity, which indicates the ability to repel water, is the most outstanding property of the rubber sheds and is what differs it from porcelain and glass. Hydrophobicity of rubber sheds assures a good electrical performance when contamination is deposited on the insulator surface, which dissolves in water and might lead to dry band arcing. Loss of hydrophobicity indicates aging of the composite insulator.

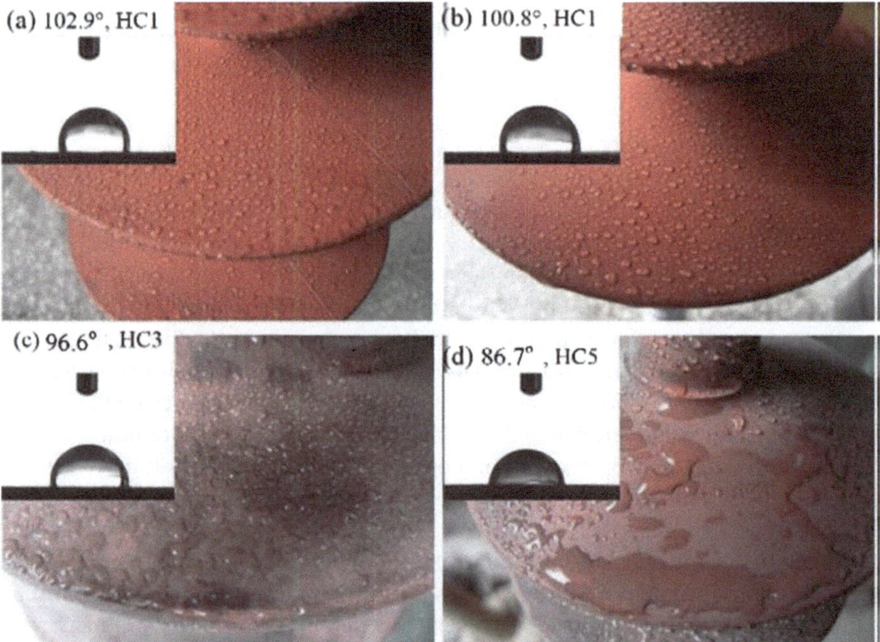

Fig. 5.3 Hydrophobicity of a composite insulator after years' operation **a** 2 years, **b** 5 years, **c** 10 years, **d** 15 years; with contact angle and hydrophobicity level in the top left corner: the bigger the contact angle, the better the hydrophobicity property the insulator has [4]

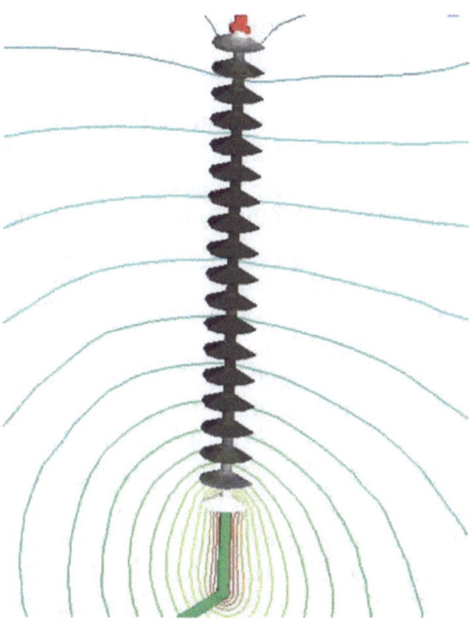

Fig. 5.4 Electric field distribution (highest magnitude in white and lowest in black) and equipotent lines around a suspension composite insulator [1]

When the rubber sheds are exposed to continues corona discharge that generates electrons, ions, ozone, ultra violet radiation, high temperature and acid, its hydrophobicity loses gradually since chemical reaction happens in the rubber with by-products of corona discharge. Figure 5.3 demonstrates the process of hydrophobicity loss on a composite insulator surface after a long-term operation [4].

When the hydrophobicity is losing, water film is easy to form on the insulator surface, as Fig. 5.3d shows. Consequently, there are more corona discharges, larger leakage current and even dry band arcing on the insulator surface. In return, the rubber degrades further and cracking and erosion may occur on the surface. Consequently, flashover happens which may lead to trip-out of the power system. For worse cases, there may be cracks generated on the surface and the Fiber Reinforced Plastic (FRP) rod is exposed to water and acid, which leads to brittle fracture of the rod and the total failure of the insulator.

5.1.3 Electric Field Distribution Around Composite Insulators

Corona discharge is always due to local electric field enhancement. Thus, the assessment of electric field magnitude distribution on an insulator surface is of great importance.

A composite insulator has a clavate shape and one end is attached by a phase conductor at high voltage potential, the electric field on the surface and in the insulator is non-uniform. Figure 5.4 shows electric field distribution on/in a suspension composite insulator and the equipotent lines around the component [1]. As it can be seen from Fig. 5.4, the highest electric field magnitude is located at the high voltage end of the insulator. While the magnitude at the low voltage end (ground potential) is also higher than that in the middle of the insulator.

There isn't any internationally accepted standard specifying the critical electric field value on the composite insulator/cross-arm surface. However, several literatures have given recommendations according to their experience, which are widely used:

Electric field magnitude, measured at 0.5 mm above the surface of the polymeric weather sheds, shall be less than 0.45 kV/mm [5]. Some literatures propose 0.42 kV/cm as the limit [6, 7]. However, these two values do not differ a lot.

It should be noticed that the critical values mentioned above, are background electric field magnitude for a clean and dry composite insulator with a hydrophobic surface.

In order to restrain the local electric field reinforcement in both ends, especially in the energized end, grading rings are stated as necessary for composite insulators at 220 kV and above, according to experiences of STRI (an independent laboratory specializing in high voltage testing in Sweden) and EPRI (Electric Power Research Institute) [6]. Figure 5.5 indicates the electric field magnitude on the surface of a composite insulator with and without a corona ring. It can be seen that the highest electric field magnitude reduces a lot when a corona ring is present, and it moves from the first shed to several sheds away along the insulator.

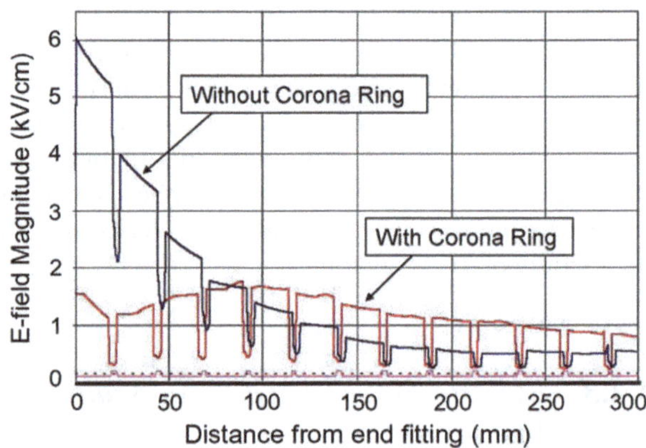

Fig. 5.5 Electric field magnitude at 0.5 mm above the sheath surface of a composite insulator for 300 mm away from the high voltage ending, with and without a corona ring [1]

5.1.4 Water Induced Corona Discharge

Corona discharge is easier to trigger if water drops are presented on the insulator surface, due to electric field concentrating locally.

Since the relative permittivity of water drops ($\varepsilon_{water} \approx 80$) is much higher than that of surrounding air ($\varepsilon_{air} \approx 1$) and rubber material ($\varepsilon_{rubber} = 2 - 10$), local electric field concentration exists around the water drops. Through numeric calculation based on Finite Element Model (FEM), Fig. 5.6 shows electric field distribution along a rubber insulator surface with a water drop presented when the electric field is parallel to the insulator surface. It can be seen that when the electric field is parallel to the insulator surface, electric field concentrates on the triple junction, where air, rubber and water meet. While, when the electric field direction is vertical to the surface with a water drop presented, the electric field concentrates at the tip of the water drop. as Fig. 5.7 shows.

The original electric field without water drops is defined as E_0, while the changed electric field magnitude after water drops presented is defined as E_{max}. Obviously in wet conditions, whether corona discharge happens or not is determined by whether the value of E_{max} exceeds the critical value E_{crit} ($E_{crit} = 26.6$ kV/cm with temperature 20 °C, pressure 101.3 kPa, and humidity 11 g/m^3 [8]) that can lead to air ionization or not. The ratio of E_{max} to E_0 is defined as electric field enhancement factor.

Many parameters have effects on the electric field enhancement factor, such as shape, volume, number, conductivity and relative location of water drops,

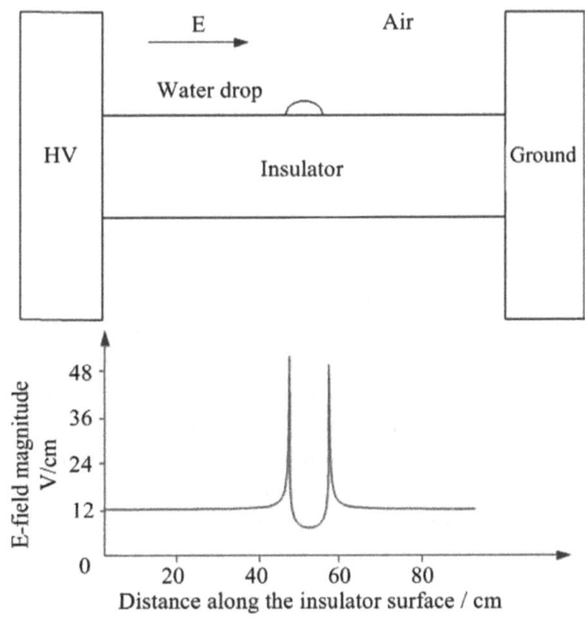

Fig. 5.6 Top: A 2-D model for numeric calculation by ANSOFT Maxwell, with an applied voltage $U = 1200$ V, gap distance between electrodes $d = 100$ cm and permittivity of air, water drop and rubber equivalent to 1, 81 and 3.7 respectively; Bottom: electric field distribution along the rubber insulator surface

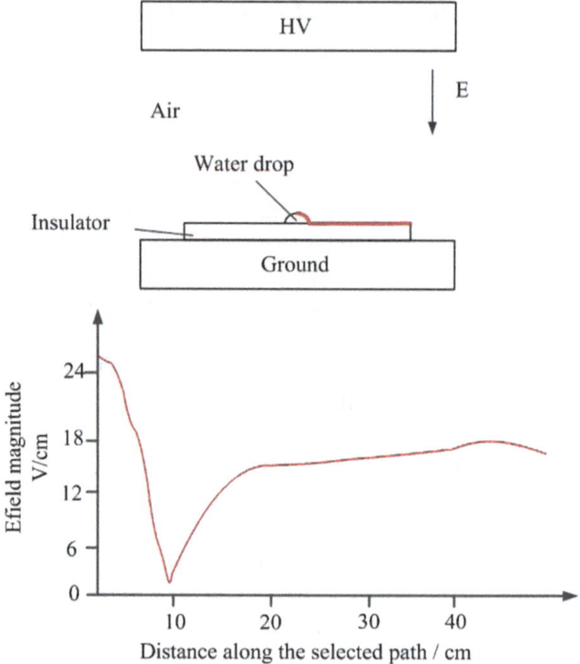

Fig. 5.7 A 2-D model for numeric calculation by ANSOFT Maxwell, with an applied voltage $U = 1200$ V, gap distance between electrodes $d = 100$ cm and permittivity of air, water drop and rubber equivalent to 1, 81 and 3.7 respectively (left); Electric field distribution along a fixed path shown in red in the 2-D model (right)

hydrophobicity level, i.e. contact angle of water drops on the polymeric material surface, etc. [9–11]. In order to investigate effects of these factors on the ratio of E_{max} to E_0, Z. Guan does a series of numeric calculation by varying parameters of water drops ($\varepsilon_{water} = 80$) on an insulator surface made from Poly Tetra Fluoro Ethylene (PTFE) ($\varepsilon_{PTFE} = 2$) [9]. Figure 5.8 indicates a qualitative relation of these factors and the ratio of E_{max} to E_0. For instance, in Fig. 5.8a, with a fixed background electric field E_0, electric field enhancement factor increases along with the volume of water drops, which means the possibility of corona discharge induced by water drops increases. Meanwhile, it is shown that E_{max} is higher with an elliptical water drop than that with a spherical water drop, especially with large water drop volume, when other conditions are identical. The similar results are obtained in [12].

It is also concluded from Fig. 5.8 that the electric field enhancement factor ranges from 2 to 8, varying according to water drop parameters and insulator surface hydrophobicity levels. In other words, occurrence of corona discharge on the surface of a polymeric insulator surface highly depends on water drop parameters and insulator surface hydrophobicity levels.

The morphology of a water drop changes when exposed to an electric field, as Fig. 5.9 shows [13]. It is obvious that the original spherical water drop elongates towards the low voltage end of the composite insulator, increasing the wet part on its surface.

Parameters of water drops deposited on the surface of a composite insulator have a stochastic nature, and an alternative electric field further increases its randomness. Thus, the value of E_{max} on a composite insulator surface is unpredictable in wet conditions.

Even though the value of E_0 is controlled as lower than the acceptable values mentioned by well-design of the weather sheds and application of grading/corona rings, the water induced corona discharge on a composite insulator surface always needs to be verified by high voltage tests.

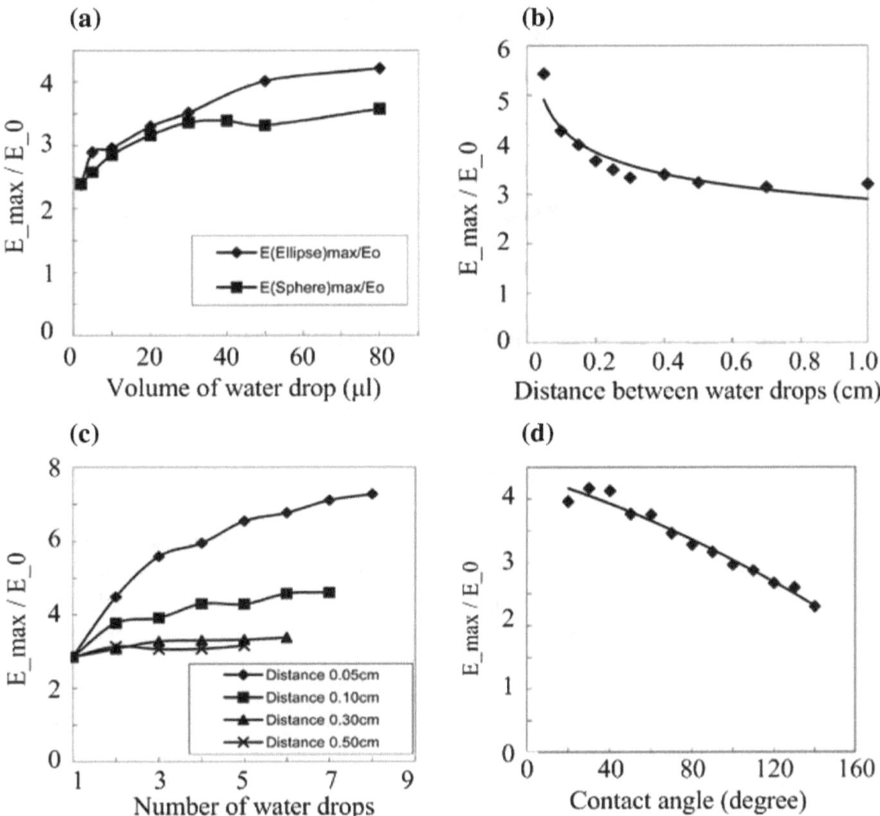

Fig. 5.8 Variation of E_{max} to E_0 according to: **a** Volume and shape of water drops; **b** Distance between water drops; **c** Number of water drops; **d** Contact angle of water drops on the polymeric materials [9]

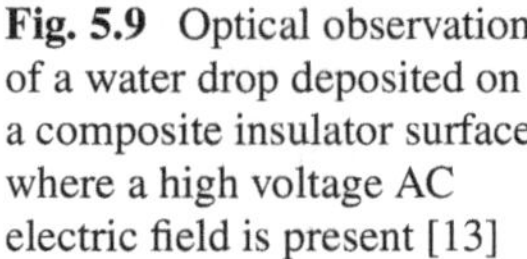

Fig. 5.9 Optical observation of a water drop deposited on a composite insulator surface where a high voltage AC electric field is present [13]

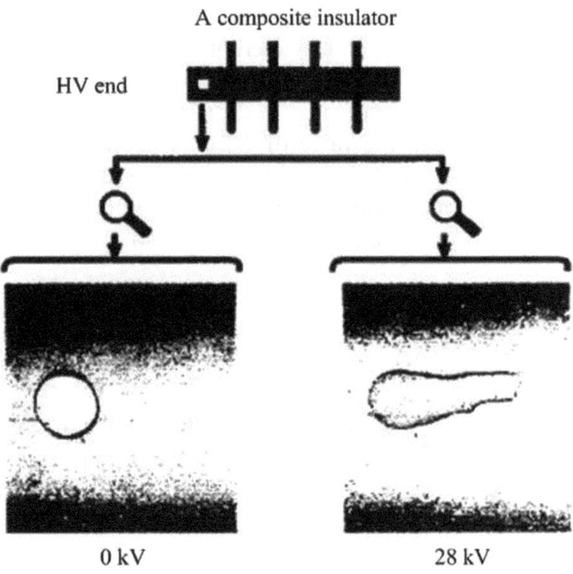

5.2 Water Induced Corona Test Circuit and Setup

In order to evaluate the electric field performance of the composite cross-arm in the fully composite pylon, a water induced corona discharge test was performed on the full-scale composite cross-arm segment in the high voltage lab of Swiss Federal Institute of Technology (ETH) Zurich, Switzerland.

The main concept of the test is to provide artificial rain to the full-scale cross-arm, apply equivalent nominal voltage to it and measure corona discharge from its surface. In order to make the experimental conditions more close to the real situation in service, the following issues should be considered in the design of test circuit and setup:

- The cross-arm needs to be hung up in the air, which is not easy considering that it has a net weight up to 281 kg.
- The inclined angle of the cross-arm from the ground plane in the test must be capable of varying in a suitable range, which requires flexible means for hanging up the cross-arm.
- An effective way of applying equivalent nominal voltage to the cross-arm must be figured out.
- An effective way of applying artificial rain to the cross-arm should be investigated.
- Corona discharges from other parts of the test setup, except for the cross-arm, must be suppressed to an acceptable level.

Considering the above-mentioned issues, the water induced corona discharge test circuit and setup have been developed.

5.2.1 Schematic of the Test Circuit

The schematic side view of the test setup for the water induced corona discharge test presented in this chapter is shown in Fig. 5.10.

The main components in Fig. 5.10 are the composite cross-arm segment, the support frame and the crane.

The composite cross-arm segment was suspended in the air. One of its end was hold by the crane fixed in the ceiling, which was able to bear a maximum vertical load of 5 tons. The other end of the segment was hanging on a metallic support frame through a strong rope and it was grounded.

The heights of the segment two ends from the ground plane were measured as h_1 and h_2. The inclined angle $\boldsymbol{\theta}_{cross-arm}$ of the cross-arm segment from the ground plane can be changed by adjusting the value of h_2 though the crane while keep $h_1 = 1.5$ m constant. Due to space limitation, a maximum value of $\boldsymbol{\theta}_{cross-arm} = 30°$ is available.

A rain simulator fixed on the ceiling was utilized to provide water drops in the test. A water pool is located below the rain simulator to collect water drops.

The rail on which the crane was fixed could be moved freely below the rain simulator. However, the crane rail must be placed next to the rain simulator. Only in this way, rain drops are not blocked by the crane rail.

It must be noted that the crane is designed for vertical loading. Thus, the hoist cable between the crane and the lifting point on the cross-arm segment must be kept perpendicular to the ground plane. Otherwise, the hoist cable may be pulled out from the winch by horizontal forces. As a result, the cross-arm may fall down, which must be avoided.

If the hoist cable of the crane hooked at the cross-arm directly, the part of the cross-arm on which the hook point was attached must be moved away from the rain simulator along with the crane. And this part cannot be exposed to rain drops, as Fig. 5.11a shows. In order to guarantee most part of the cross-arm segment, especially the part at high voltage potential, could be exposed to rain drops, the cross-arm segment was 'elongated' by an aluminum square rod, as Fig. 5.11b indicates. In this case, the hoist cable of the crane hooked at the extension rod, instead of at the cross-arm segment directly. Resultantly, the most part of the cross-arm can be exposed to rain drops.

A conventional suspension insulator for 400 kV lines was applied between the extension rod and the crane. To restrain corona discharges from the metallic rod and the insulator, four corona rings are utilized. Meanwhile, the nearest corona ring from the cross-arm segment, i.e. corona ring 1, also worked as high voltage electrode.

The extension rod was connected to a 800 kV AC transformer. Also, it was connected to a coupling capacitor for partial discharge (PD) measurement.

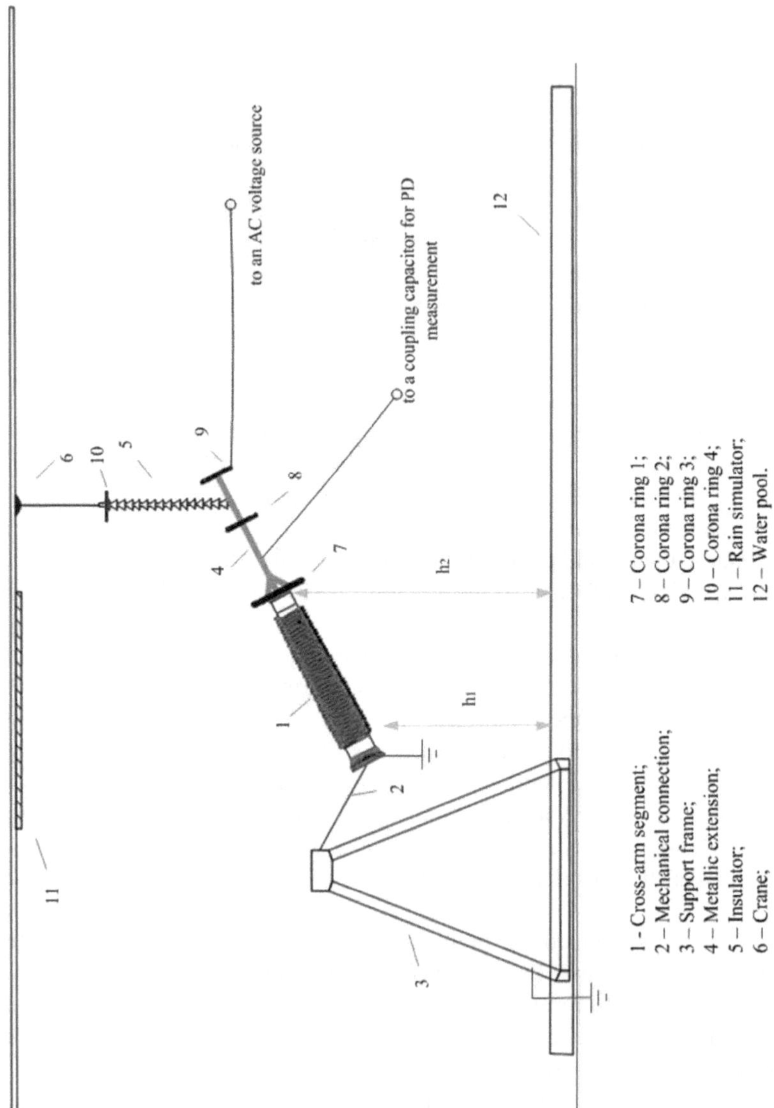

Fig. 5.10 Schematic side view of water induced corona test circuit

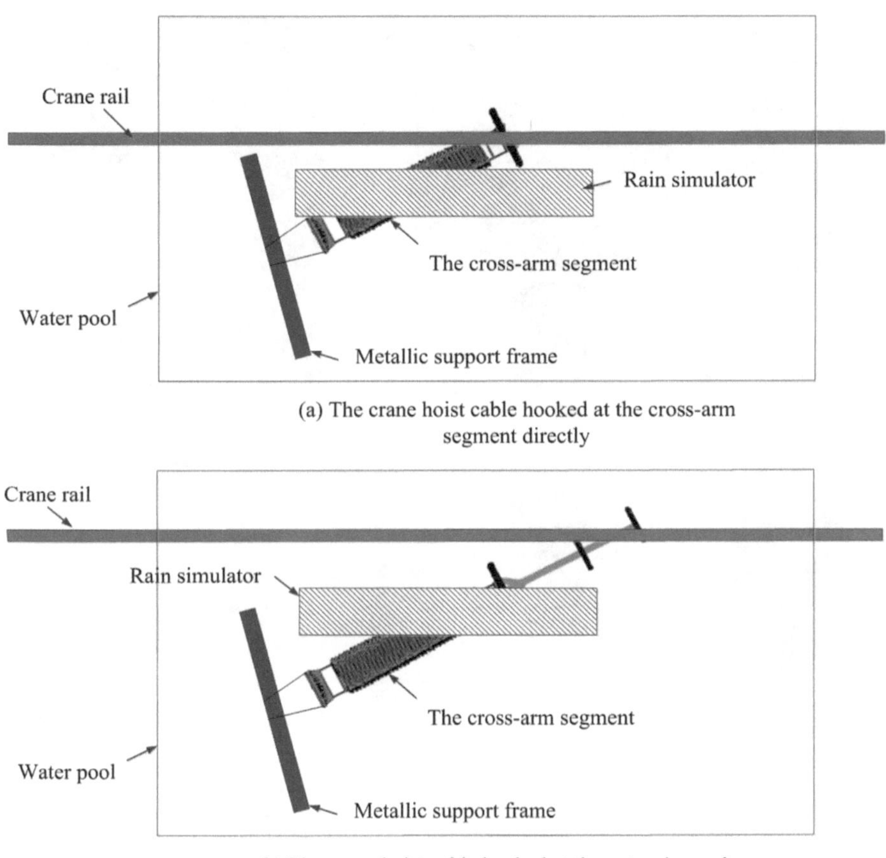

Fig. 5.11 Sketch map from top view of relative locations of critical components in the water induced corona test circuit: **a** without the extension rod; **b** with the extension rod

5.2.2 Test Setup

The setup arrangement for the water induced corona discharge test is shown in Fig. 5.12.

The critical components of the setup will be introduced as following.

5.2.2.1 The Composite Cross-Arm Segment

It should be noted that, the final manufacture of the composite cross-arm for the fully composite pylon takes quite a long time and has not been finished yet. However, it is found out that a conventional composite bushing for high voltage application provided by ABB, Sweden, has the same structure and similar configuration with

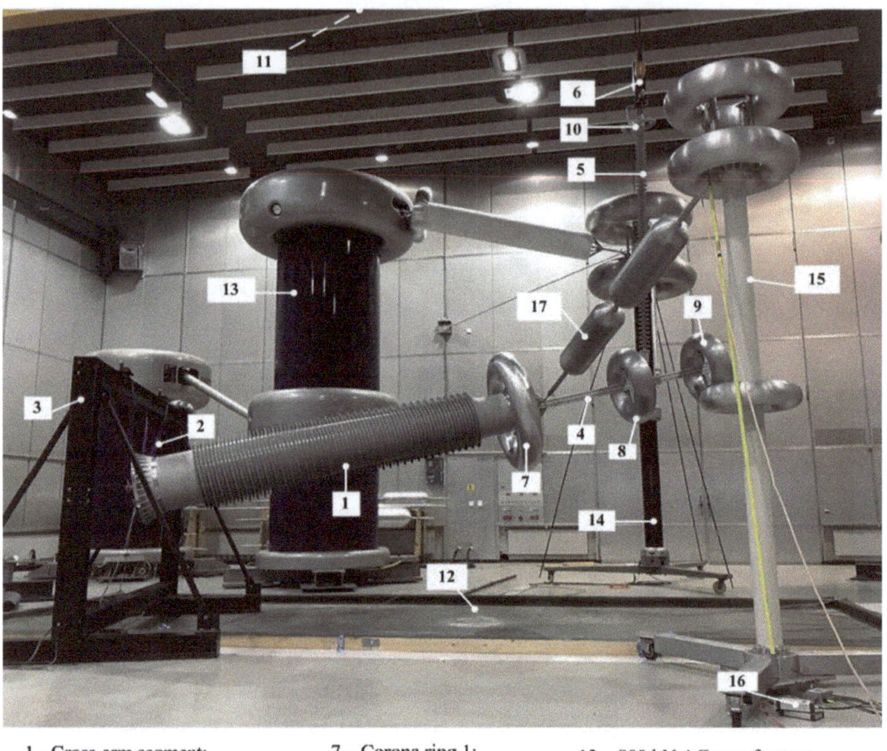

1 - Cross-arm segment;
2 – Mechanical connection;
3 – Support frame;
4 – Metallic extension;
5 – Insulator;
6 – Crane;
7 – Corona ring 1;
8 – Corona ring 2;
9 – Corona ring 3;
10 – Corona ring 4;
11 – Rain simulator;
12 – Water pool;
13 – 800 kV AC transformer;
14 – Voltage divider;
15 – Coupling capacitor;
16 – OMICRON MPD600
17 – HV connection

Fig. 5.12 Experimental setup for water induced corona test

that of the composite cross-arm in the fully composite pylon. Analysis and tests on the composite bushing are valuable references for the composite cross-arm. Thus, in the present section, the composite bushing is regarded the same as the composite cross-arm segment.

The cross-arm segment presents a section between the top phase and middle phase in the fully composite pylon. The section in the pylon is stressed by a phase-to-phase voltage. Figure 5.13 displays the configuration of the cross-arm segment.

The segment is comprised of a hollow conical Fiber Reinforced Plastic (FRP) tube made from fiberglass reinforced epoxy, weather sheds/sheath made from High Temperature Vaculized (HTV) silicon rubber and two metallic flanges for both ends.

The High Temperature Vaculized (HTV) silicon rubber material has a HC1 hydrophobicity level (the highest level for hydrophobicity), and is also considered

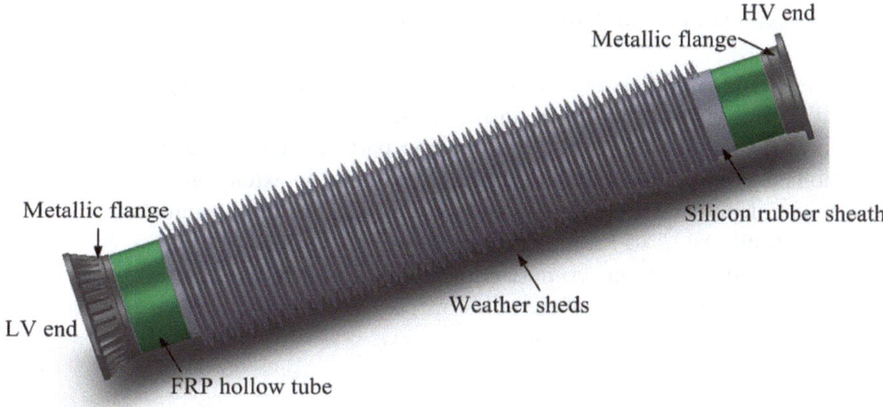

Fig. 5.13 The full-scale composite cross-arm segment

as a Hydrophobicity Transfer Material (HTM). Thus, it's not easy for water droplets forming on the silicon rubber surface, which restrains the corona discharge.

The weather sheds of the cross-arm segment adopts alternative profiles. All parameters of the sheds fulfill the recommendations for insulators by IEC Standard 60815 [14].

For better expression, the smaller end of the segment is defined as high voltage end, while the bigger end is defined as low voltage end. At the high voltage end, the silicon rubber sheath has an extension along the bare Fiber Reinforced Plastic (FRP) tube by 140 mm. Thus between the silicon rubber sheath and the high voltage flange, there is a section of Fiber Reinforced Plastic (FRP) bare tube with a length of 250 mm. The dimensions of the cross-arm segment is shown in Table 5.1.

Since the drawing of the composite cross-arm must be kept confidential as required by the supplier, more details regarding the cross-arm segment cannot be presented here.

Table 5.1 Dimensions of the cross-arm segment in mm

Total length	3640
Minimum arcing distance	3365
Silicon rubber length	2806
Creepage distance	11200
Diameter of the fiber reinforced plastic (FRP) tube	385/510[a]
Diameter of the flanges	515/655[b]

[a]385 for the high voltage end, while 510 for the low voltage end;
[b]515 for the high voltage flange, while 655 for the low voltage flange

5.2.2.2 Extension of the Cross-Arm Segment

As discussed in Sect. 5.2.1, an extension of the cross-arm is necessary to expose most part of the cross-arm to rain drops during tests.

Figure 5.14 shows details of the extension part of the cross-arm segment. The extension part was comprised of a connection part and an extension square rod. Both the connection part and the extension rod were made from the aluminum construction profiles—Profile 80 × 80 Nature-produced by ITEM [15].

The connection part was comprised of a cross, which was screwed to the high voltage (HV) metallic flange of the cross-arm segment, and 4 short support bars that

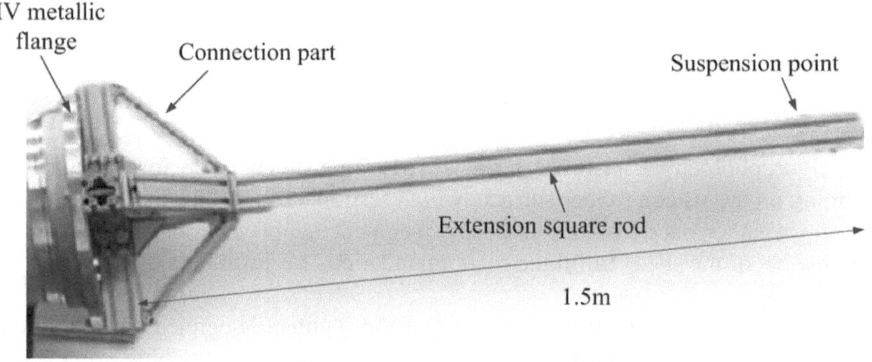

(a) Entire side view of the extension part of the cross-arm segment

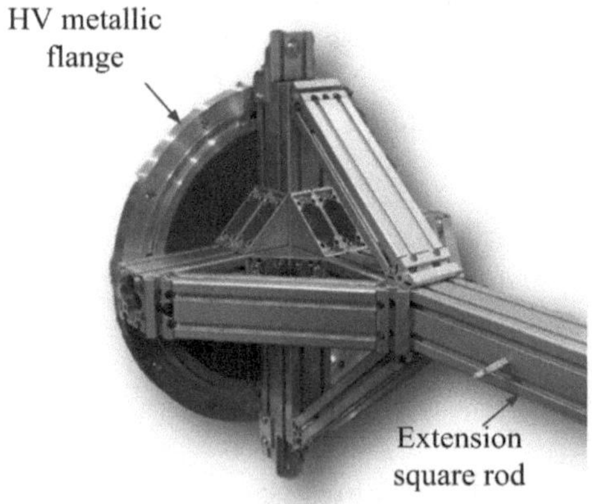

(b) Enlarged view of the connection part between the cross-arm segment and the extension rod

Fig. 5.14 Extension part of the cross-arm segment

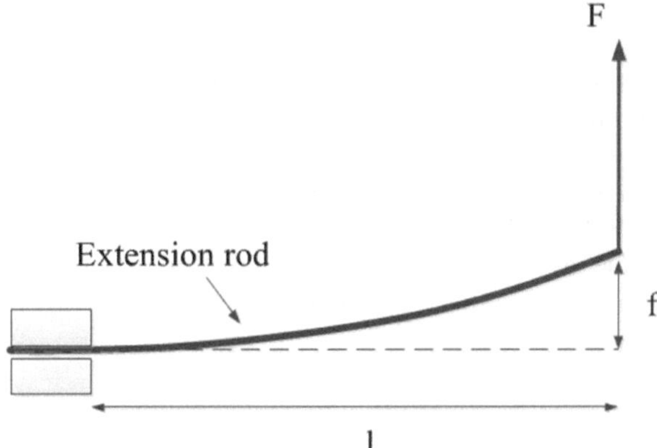

Fig. 5.15 Extension part of the cross-arm segment

are utilized to sustain the bending moment of the extension rod. The extension rod with a length of 1.5 m was screwed to the cross center.

The square rod need to carry the weight of the cross-arm segment and its dead weight including all the other accessories. The extension rod can be regarded as a cantilever beam and the load scenario in the present case is shown in Fig. 5.15. F represents the total amount of dead load in N, accounting all parts which were lifted by the crane. $l = 1500$ mm represents the length of the rod, while f represents the maximum deflection of the extension rod in mm. The net weight of the cross-arm segment is 281 kg and the total weight that the rod needs to carry is considered as 350 kg. Thus, the value of F can be calculated as $F = 350\,\text{kg} \times 10\,\text{N/kg} = 3500\,\text{N}$. The maximum deflection of the extension rod can be calculated as [16]

$$f = \frac{F \times l^3}{3 \times E \times I} = 30.31\text{mm} \tag{5.1}$$

where $E = 70000\ \text{N/mm}^2$ is the modulus of elasticity of the rod, while $I = 187.7\ \text{cm}^4$ is the moment of inertia.

It can be seen that the maximum deflection is not remarkable compared with the setup dimensions, thus it's acceptable.

In order to make sure the rod will not be deformed permanently, the bending stress σ of the rod when carrying all the weights is calculated by [16]

$$\sigma = \frac{F \times l}{I/t} = 111.88\ \text{N/mm}^2 \tag{5.2}$$

where $t = 40$ mm is half the thickness of the extension rod.

The yield point of the extension rod is 195 N/mm^2 [15], which is larger than σ. Thus, the extension rod is able to lift up the cross-arm segment as design.

5.2.2.3 The Metallic Support Frame

Figure 5.16 displays the metallic support frame from which the low voltage end of the cross-arm segment was hanging. The dimensions of the frame are shown in the figure. The frame was made from steel thus it was quite strong and stable, which could bear the mechanical load of the cross-arm segment.

It should be noted that a small corona ring was applied at the top left corner of the frame, as shown in Fig. 5.16. It aimed to restrain corona discharge from this sharp corner, which stood quite close to the AC transformer.

5.2.2.4 Corona Rings

As shown in Fig. 5.12, a total number of 4 corona rings were applied in the setup.

Corona ring 1, a toroid made from aluminum, was applied at the high voltage end of the cross-arm segment, as Fig. 5.17 shows. There were two functions of the Corona ring 1: (1) introduce high voltage to the cross-arm segment; (2) restrain corona discharge from the high voltage metallic flange and extension connection part. The corona ring has an internal and external diameter of 660 mm and 1100 mm, respectively. The ring was fixed on the metallic flange through two aluminum square rods. The square rods were also made from Profile 80 × 80 Nature produced by ITEM [15].

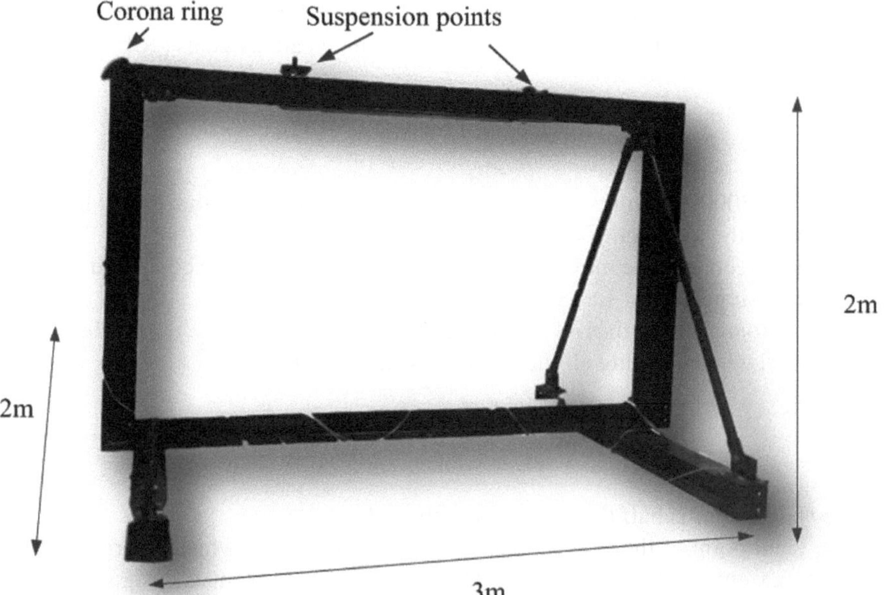

Fig. 5.16 Metallic support frame on which the cross-arm segment was hanging

(a) The cross-arm segment with the corona ring

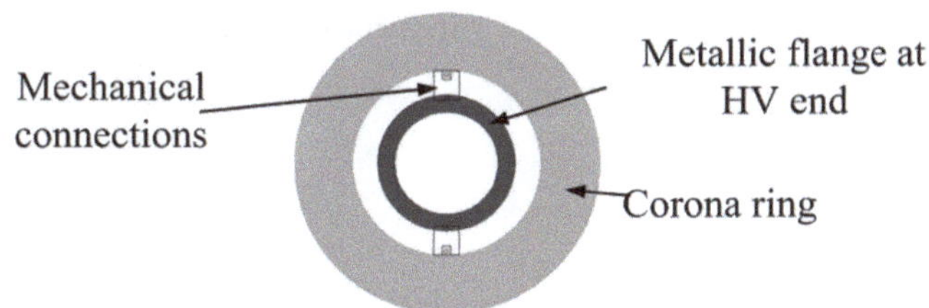

(b) Lateral view of the cross-arm segment with the corona ring

Fig. 5.17 Corona ring 1 on the full-scale cross-arm segment

Corona ring 2 and corona ring 3 were also toroids made from aluminum and they had the same dimensions. They had an internal and external diameter of 360 mm and 560 mm, respectively. The two corona rings were fixed by two aluminum square bars on the extension rod with a distance of 500 mm, as Fig. 5.18 shows.

Figure 5.19 displays the corona ring 4 which was applied to restrain corona discharge from the suspension insulator between the crane and the extension rod.

5.2.2.5 The High Voltage Connections

In order to restrain corona discharge from connections between the transformer, cross-arm segment and the coupling capacitor, a special high voltage connection rod were applied which was thicker in diameter than normal high voltage cables. Figure 5.20 indicates the shape of the connection rod.

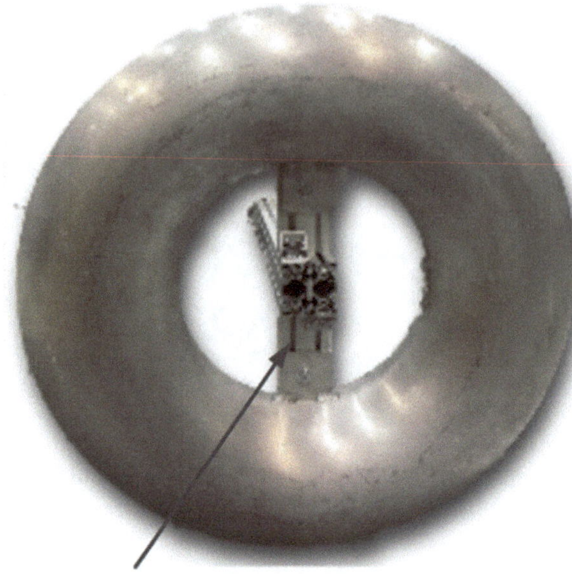

Fig. 5.18 Arrangement of corona ring 2 and corona ring 3

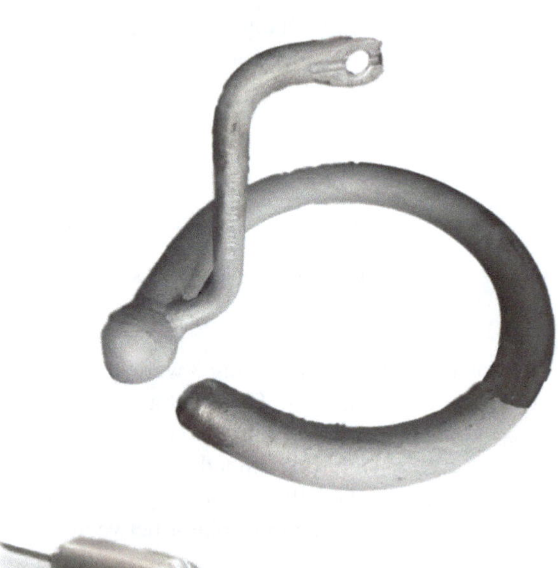

Fig. 5.19 Corona ring 4

Fig. 5.20 The high voltage connection rod

The rod was made from aluminum and contained two enlarged cylinders in the middle that were critical for corona suppression.

5.2.2.6 The Rain Simulator

The rain simulator is an array consisted of approximate 500 hollow needles and located in the ceiling with a height of 10 m. It has a width of 0.6 m and a length of 4 m. Figure 5.21 shows a part of the rain simulator.

The needle array was connected to a tap water supply, which was controlled by a water valve. The water flow rate was recorded by LABJACK, a measurement device which transfered analog water flow signals into digital signals.

In order to transfer the water flow into rain density, a calibration process was applied. The rain simulator was kept on for 30 min and the flow rate was controlled at a constant value v_0. During this period, 5 boxes were put in a line along the length of the rain simulator array at a distance of 30 cm and water were collected by these boxes. The base area of the box was recorded as s in the unit of mm^2 and water volume collected by the 5 boxes were recorded as V_1-V_5 respectively in the unit of mm^3. Then the rain density RD_0 could be calculated by

$$RD_0 = \frac{V_1 + V_2 + V_3 + V_4 + V_5}{5} * \frac{1}{s} * \frac{30}{60} \text{ mm/h} \tag{5.3}$$

Then the scale factor S_{f-RD} to transfer the water flow rate into rain density can be obtained by

$$S_{f-RD} = \frac{RD_0}{v_0} \tag{5.4}$$

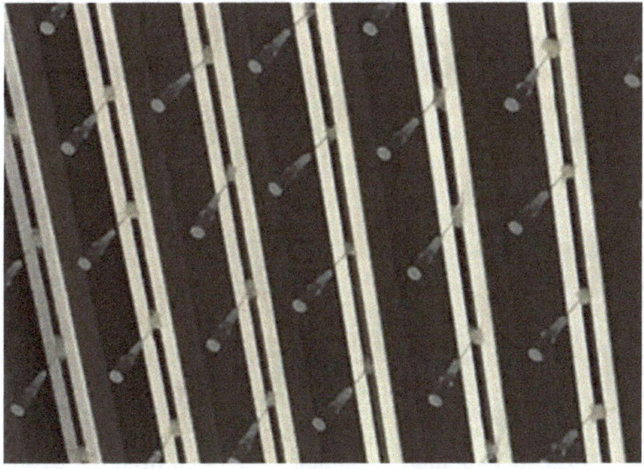

Fig. 5.21 A part of the rain simulator array

As a result, any water flow rate v_x can be transfered into ran density RD_x by

$$RD_x = v_x * S_{f-RD} \tag{5.5}$$

5.3 Electric Field Distribution on the the Composite Cross-Arm

It's impossible to compute the electric field magnitude on the cross-arm surface in wet conditions accurately, since the randomly distributed water droplets have great effects on the results. Instead, the electric field distribution on and around the cross-arm surface in dry conditions is commutated as a reference for the prediction of corona discharge activities in wet conditions.

5.3.1 Electric Field on the Cross-Arm Surface with Initial Design

In order to analyze the electric field distribution on and around the surface of the composite cross-arm, a 2D electric field computation has been performed based on the initial design of the fully composite pylon and the results are displayed in [17].

For the voltage loading, the worst case has been considered, i.e. the upper phases are loaded with U_U, the peak value of the highest system voltage 420 kV, while a negative voltage U_M and U_L with half of the peak value with a phase angle of $\pm 120°$ are loaded to the middle phases and lower phases respectively.

More details regarding parameters for the electric field computation are referred to [17]. According to Sect. 5.1.3, the critical region for assessing the electric field magnitude on the composite insulator surface should be at 0.5 mm above the weather sheath. Figure 5.22 displays the electric field magnitude at 0.5 mm from the weather sheath of the composite cross-arm section between the upper and middle phase in the fully composite pylon.

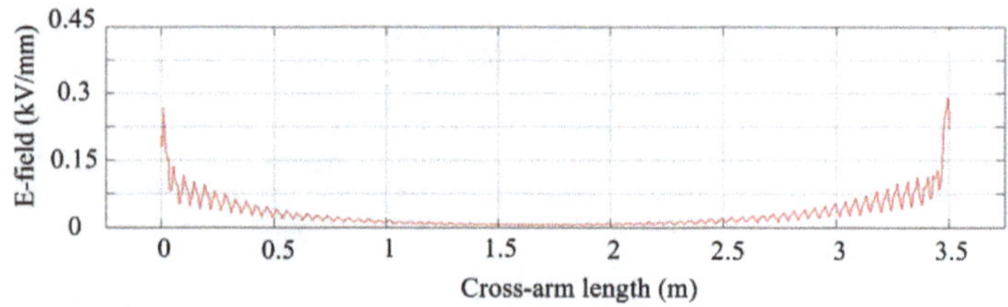

Fig. 5.22 Electric field magnitude at 0.5 mm from the weather sheath surface of the composite cross-arm with the initial design [17]

It can be seen from Fig. 5.22 that the electric field at both ends of the cross-arm section intensifies, where phase conductors are located. While the electric field magnitude is the lowest in the middle of the cross-arm section.

The calculated peak value E_{max} of the electric field at 0.5 mm from the weather sheath is 0.283 kV/mm.

The maximum value is lower than the maximum acceptable electric field magnitude (0.45 kV_{rms}/mm) at 0.5 mm from the weather sheath, therefore theoretically, corona discharge is unexpected to observe from weather sheds in wet conditions. However, this conclusion needs to be experimentally investigated.

5.3.2 Electric Field on the Cross-Arm Segment in the Test

As introduced in Sect. 5.2.2.1, the cross-arm segment applied in the test was an alternative for the initial designed cross-arm section in the full composite pylon. In order to make the test results comparable with the real situation in the fully composite pylon, electric field distribution on and around the cross-arm segment in the test was investigated.

Since a phase-to-phase voltage was not available in the high voltage lab, an equivalent phase-to-ground voltage was applied. A sinusoidal voltage with a peak value of U_{equ} was applied in the test. U_{equ} was obtained by

$$U_{equ} = 420 \times \sqrt{2} = 594 \text{ kV} \tag{5.6}$$

A 2D electric field computation based on ANSOFT MAXWELL was performed, with a sinusoidal voltage with a peak value of 594 kV applied at the corona ring 1 and the high voltage metallic flange, while the low voltage metallic flange was at zero potential. The relative permitivity of the Fiber Reinforced Plastic (FRP) tube, silicon rubber sheds and the air were 5.75, 3.67 and 1 respectively. The electric field magnitude at 0.5 mm from the weather sheath was shown in Fig. 5.23.

The maximum electric field magnitude was measured at edge of the last weather shed in the vicinity of the high voltage end, which was expected. This maximum value was 0.273 kV/mm. Compared with the maximum electric field magnitude in the case of the cross-arm with initial design (0.283 kV/mm), the two maximum values are identical approximately.

Thus, it proves that the test on the cross-arm segment can be considered as valuable reference for test on the real cross-arm in the 'Y' pylon. Also, the equivalent phase-to-ground voltage is reasonable in the water induced corona discharge test.

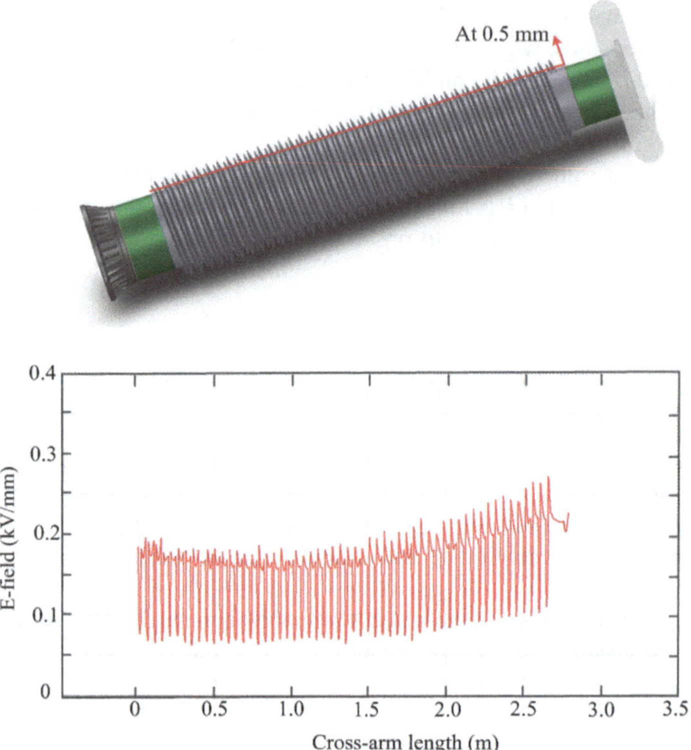

Fig. 5.23 Electric field magnitude at 0.5 mm from the weather sheath surface of the composite cross-arm segment in the test

5.4 Water Induced Corona Discharge Test

5.4.1 Test Procedure

Before voltage was applied, the rain simulator was turned on and tap water was applied to the test setup. Water flow rate was controlled at a constant value. According to calibration, the water flow rate was equaled to a constant rain density $RD = 7.3$ mm/h, which indicated a light to moderate rain density [18]. After the rain density was stable, the rain was kept on for 30 min and then turned off.

Since only the corona activities from the silicon rubber was of interest to us, after rain, water droplets on the metallic flanges, bare composite tube, corona rings and other parts of the setup were wiped dry to restrain corona activities from these parts.

After that, step up the applied voltage V_0 from 0 to 75% of the target value 420 kV_{rms} at a rate of 10 kV/s. Then step up it continuously at a rate of 5 kV/s, until it reached the target voltage. The applied voltage was kept as $V_0 = 420\ \text{kV}_{\text{rms}}$ for 5 min. Afterwards, the applied voltage was decreased to 0 kV.

During the whole voltage application period, corona discharge signals were measured by the partial discharge (PD) measurement unit OMICRON MPD 600. The partial discharge (PD) measurement unit was set in accordance with requirements of IEC Standard 60270 [19]. A center frequency $f_m = 180$ kHz and a bandwidth of 160 kHz were chosen for filtering. The corona measurement threshold was set at 10 pC. The voltage at which a corona discharge signal was measured for the first time was recorded as corona inception voltage. The phase resolved partial discharge (PRPD) patterns, partial discharge (PD) peak magnitude, partial discharge (PD) average magnitude and partial discharge (PD) intensity (repetitive rate) were measured.

After the corona discharge inception voltage was measured, the cross-arm segment was left for drying out at room temperature. This process took around 24 h. After the cross-arm was totally dry, the above test process was repeated and another corona discharge inception voltage was recorded. The test was performed 3 times in total and the average value of corona inception voltage was calculated as V_{inc}, based on the three measurements.

Also, during the voltage application period, corona discharges on the setup were observed by the experimental operating personnel through a corona camera - CoronaFinder made by SYNTRONICS with a sensitivity of 10 pC.

In order to capture possible corona discharge phenomenon, a digital camera was applied, which was located behind the corona camera thus corona discharge photos were shot through the corona camera lens.

For comparison, the above procedures were also performed on the cross-arm segment in dry conditions.

The ambient atmosphere conditions were: (1) temperature: 21.4 °C; (2) relative humidity: 41.8%; (3) pressure: 95.36 kPa.

5.4.2 Test Results

Test results are shown as following.

5.4.2.1 Corona Discharge Inception Voltage and Electric Field Magnitude

In dry conditions, no corona activities were observed by the corona camera, nor measured by the partial discharge (PD) measurement unit. However, with water droplets presented, corona activities were observed and measured.

With three measurements, an average corona inception voltage V_{inc} with a value of 379 kV_{rms} was obtained.

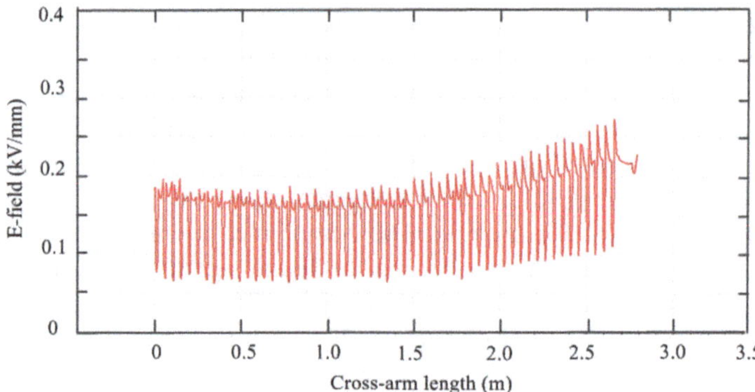

Fig. 5.24 Electric field distribution at 0.5 mm from the weather sheath with a voltage of 536 kV with which corona discharge was initiated

The maximum background electric field magnitude was computed on the cross-arm segment surface, when a corona discharge was triggered. The voltage of $379 \times \sqrt{2} = 536$ kV was applied to the 2D electric field computation program, which was introduced in Sect. 5.3.2. As a result, the electric field distribution at 0.5 mm from the weather sheath was shown in Fig. 5.24.

In can be seen from Fig. 5.24 that the maximum background electric field at 0.5 mm from the weather sheath was 0.256 kV/mm, when corona discharge was induced by water droplets. The computed value is lower than the maximum electric field magnitude 0.283 kV/mm with nominal voltage in the system. Meanwhile, it's lower than the maximum acceptable value 0.45 kV/mm.

5.4.2.2 Corona Discharge Characteristics

Figure 5.25 shows phase resolved partial discharge (PRPD) patterns in dry and wet conditions.

It can be seen from Fig. 5.25 that in dry conditions, a partial discharge (PD) signal with a level of 10 pC approximately labeled as '1' was measured. According to shape of the PD signal '1', it was regarded as noise from the setup. While in wet conditions, another partial discharge (PD) signal labeled as '2' was measured. PD signal 2 was a typical surface discharge, which was regarded as corona discharge due to water droplets on weather sheds surface.

With an applied voltage $V = 420\ \text{kV}_{\text{rms}}$ for 5 min, the partial discharge (PD) peak magnitude from the cross-arm segment surface was 834 pC, while the average partial discharge (PD) level was 80 pC. The partial discharge (PD) repetitive rate was measured as 52 PDs/s.

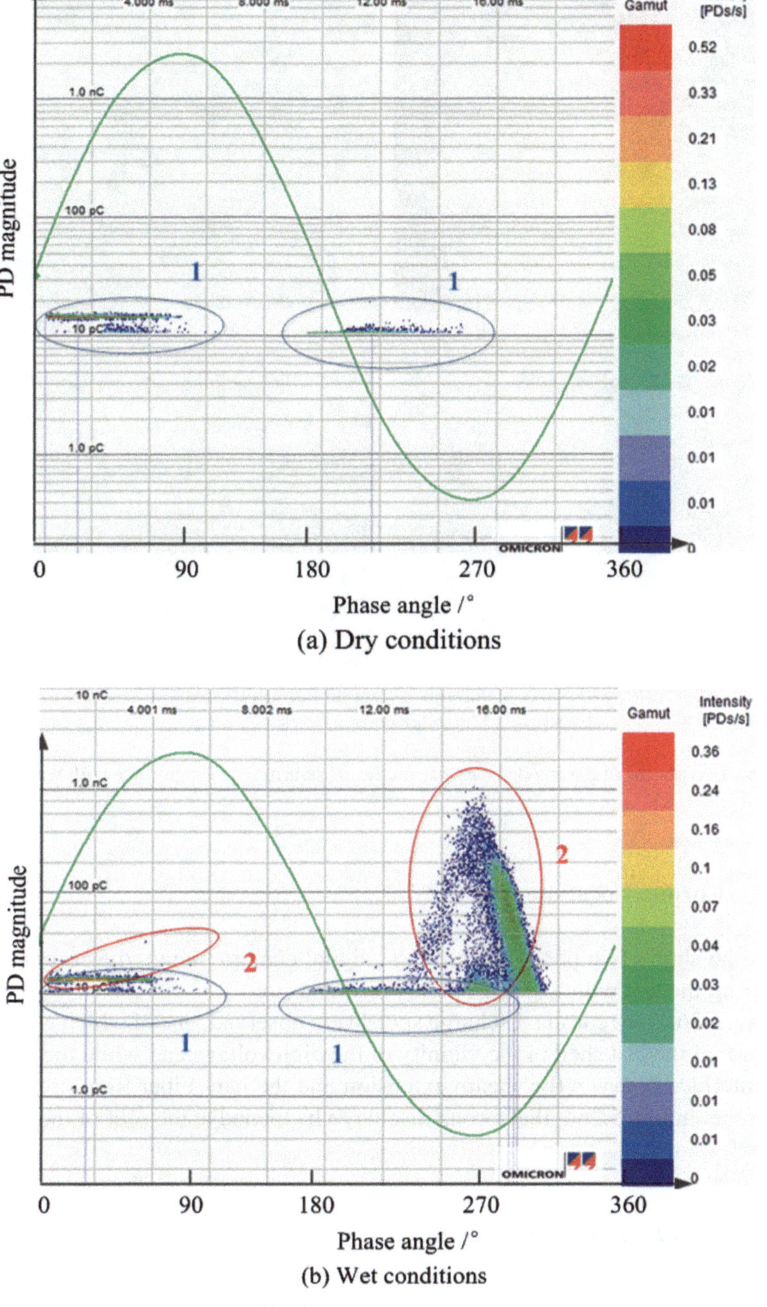

Fig. 5.25 Partial discharge (PD) levels versus applied voltage phase angle in dry and wet conditions

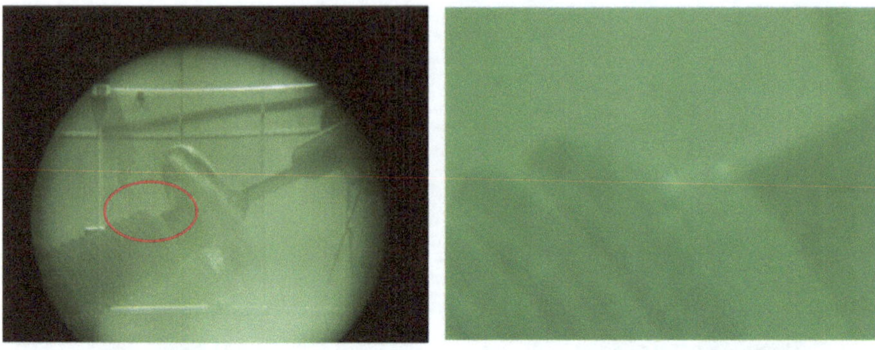

(a) Corona discharge with weak light (left); zoom- in view of corona discharge (right).

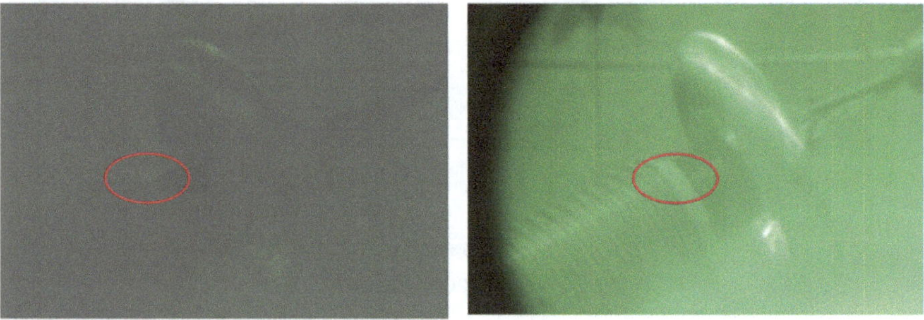

(b) Corona discharge without background light (left); photo of setup with background light for corona discharge location.

Fig. 5.26 Two photos of corona discharge from the cross-arm segment surface with water droplets presented at 420 kV_{rms}

5.4.2.3 Corona Locations

Figure 5.26 shows two photos shot by the digital camera though the corona camera lens during the test with water droplets presented.

In Fig. 5.26a, two corona discharge spots were observed, one of which was located at the root of the last shed in the vicinity of the high voltage end while the other was at the interface between the sheath extension and the bare Fiber Reinforced Plastic (FRP) tube. In Fig. 5.26b, the corona discharge happened at the root of the last shed.

5.4.3 Effects of Inclined Angles

Since the inclined angle of the cross-arm from the ground plane have effects on the water droplet's location, number, size and shape deposited on its surface, corona discharge from the cross-arm surface may be affected by its inclined angle.

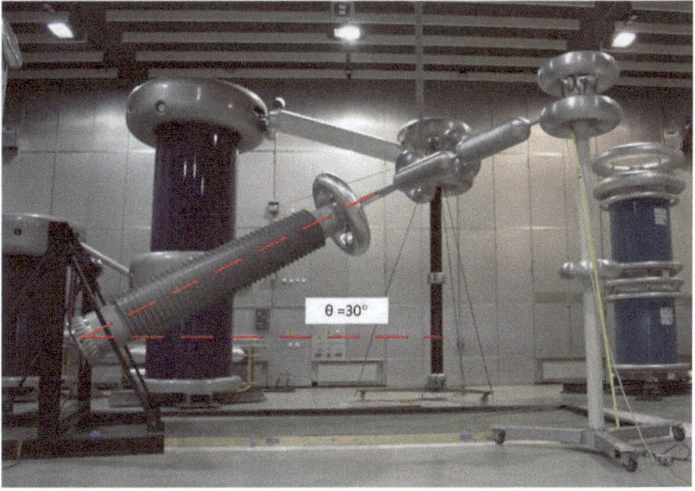

Fig. 5.27 Setup with different cross-arm segment inclined angles in the water induced corona discharge test

In order to investigate effects of inclined angles on the water induced corona discharge from the cross-arm surface and find out the optimal inclined angle, the water induced corona discharge tests were performed with 5 cross-arm inclined angles, from 10° to 30° at a step of 5°. Figure 5.27 indicates different arrangement of the setup with the smallest and the biggest cross-arm inclined angles.

5.4.3.1 Test Process

The same process was performed as introduced in Sect. 5.4.1. During the rain exposure period, shape, size and location of water droplets on the cross-arm silicon rubber surface were recorded by a digital camera Nikon D5300.

5.4.3.2 Test Results

Figure 5.28 indicates measured partial discharge (PD) characteristics with different inclined angles, where error bars represent standard deviations.

For cases with a cross-arm inclined angle of 10°, 15° and 20°, the partial discharge (PD) inception voltage was almost the same −420 kV_{rms}. While, with an increased cross-arm inclined angle of 25° and 30°, the partial discharge (PD) inception voltage was reduced to 391 kV_{rms} and 379 kV_{rms} respectively.

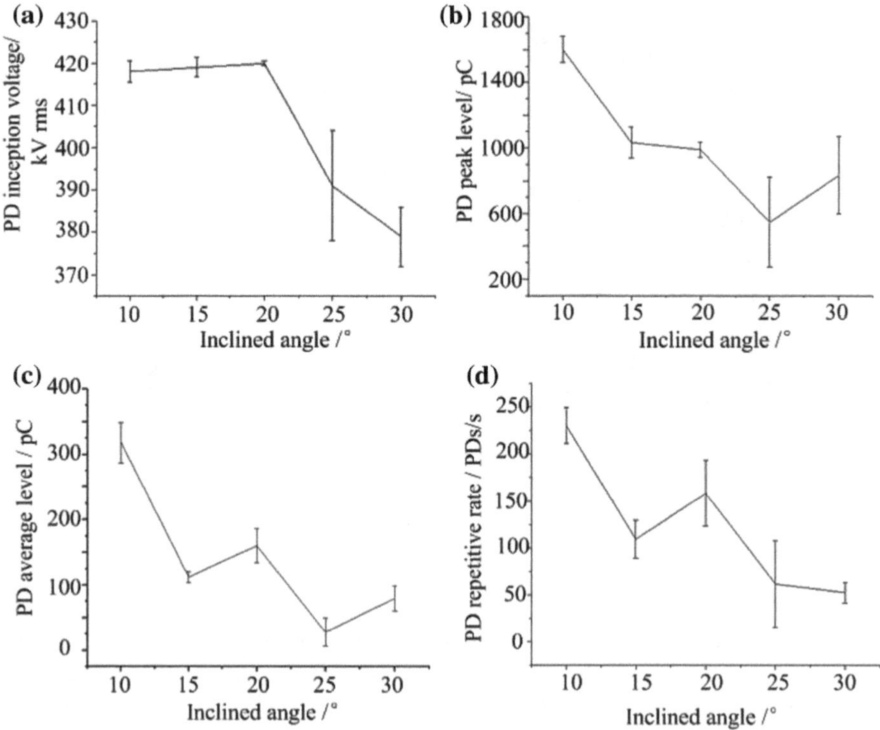

Fig. 5.28 Different cross-arm segment inclined angles versus: **a** partial discharge (PD) inception voltage; **b** partial discharge (PD) peak level; **c** partial discharge (PD) average level; **d** partial discharge (PD) repetitive rate

For other partial discharge (PD) characteristics, i.e. partial discharge (PD) peak level, partial discharge (PD) average level and partial discharge (PD) repetitive rate, with increase of the inclined angle, the measured partial discharge (PD) values had a trend of decreasing approximately, although they were not linear. The reason for the phenomenon may be that with the increase of inclined angle, the number of water droplets, which can deposit on the weather sheath, is decreasing. With less water droplets, the partial discharge (PD) peak level, partial discharge (PD) average level

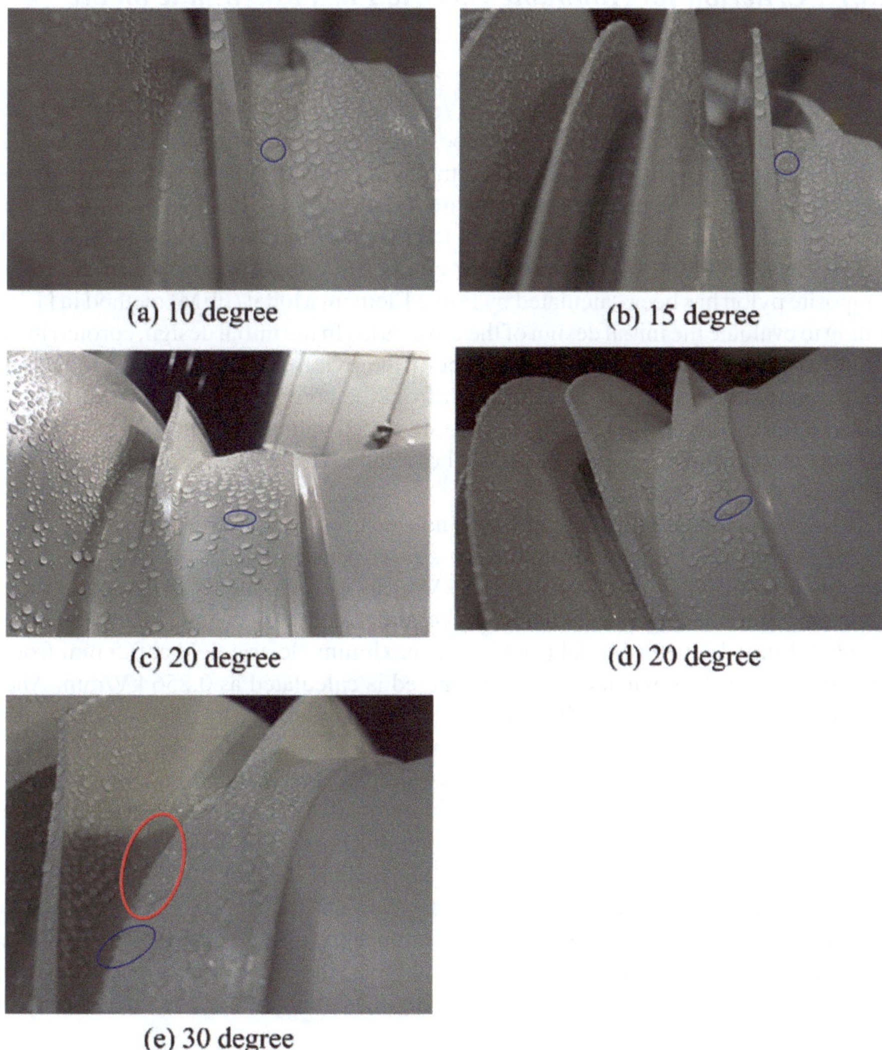

Fig. 5.29 Photos of water droplets morphology near shed root with various cross-arm inclined angles

and repetitive rate are reduced. However, photos in Fig. 5.29 does not show obvious decrease of water droplets number. The reason for the partial discharge (PD) values decreasing with increasing of inclined angles needs further research in the future.

5.5 Discussion

5.5.1 *Criterion for Allowable Electric Field Magnitude on the Cross-Arm Surface*

Some literatures recommend 0.45 kV_{rms}/mm as the maximum allowable electric field magnitude at 0.5 mm from the composite insulator/cross-arm surface [17, 20]. If this criterion is fulfilled, it is believed that no corona discharge will be triggered from the composite insulator/cross-arm in wet conditions. It should be noted the value of 0.45 kV_{rms}/mm is defined when the insulator/cross-arm is clean and dry.

The electric field at 0.5 mm from the composite cross-arm surface in the fully composite pylon has been calculated by Finite Element Model (FEM) method in [17], aiming to evaluate the initial design of the cross-arm (In the initial design, corona rings are not considered for an aesthetic impact of the pylon). The calculation indicates that the maximum electric field at 0.5 mm from the composite cross-arm surface is 0.283 kV/mm, which is lower that the recommended value 0.45 kV_{rms}/mm. Thus, it is rational to expect that no corona should be observed from the composite cross-arm surface in wet conditions.

However, in the present chapter, water induced corona discharge test displays that corona discharge is triggered from the cross-arm surface ($\boldsymbol{\theta}_{cross-arm} = 30°$) with an equivalent phase-to-ground voltage 379 kV_{rms} in wet conditions. By applying the corona inception voltage as the boundary condition in the electric field computation based on Finite Element Model (FEM), the maximum electric field at 0.5 mm from the cross-arm surface when corona is triggered is calculated as 0.256 kV/mm. And the rain density considered in the test is 7.3 mm/h, defined as a light to moderate rain in [18].

It is obvious that a deviation exists between the recommended maximum allowable electric field magnitude on the composite insulator/cross-arm surface and the experimental electric field with which corona discharge is initiated from the composite cross-arm surface in wet conditions. The main reason for this deviation is that the corona detection criteria in these two cases are different.

When developing the recommended maximum allowable electric field magnitude, researchers consider the visible corona discharge as the criterion, i.e. only when the corona discharge is visible to naked eyes or digital camera, it is recorded as a corona initiation [7, 17, 20].

While in the present chapter, the corona discharge is monitored by a partial discharge (PD) measuring unit with a threshold of 10 pC, i.e. as long as there is a partial discharge (PD) signal with a magnitude higher than 10 pC, it is registered as a corona initiation. For sure a partial discharge (PD) signal with a magnitude around 10 pC is invisible to naked eyes or normal digital cameras. And it is the truth that in the water induced corona discharge test, no corona discharges have been captured by the digital camera directly. Instead, photos, shown in Fig. 5.26, displaying corona from the cross-arm segment surface were captured through a corona camera with a sensitivity of 10 pC. It can be seen from Fig. 5.26 that the corona are too weak to be captured by naked eyes. Thus, the criterion applied in the present chapter is more stringent, compared with the recommended values.

One feasible way to compare the experimental corona inception electric field with the recommended allowable maximum electric field magnitude is to use the visible corona as the corona initiation criterion in the test. In this case, the applied voltage in the test must be increased. A higher applied voltage requires a longer suspension insulator to isolate the crane. Due to space limitation, with the present test setup, a higher applied voltage is not available.

For the electrical design of the fully composite pylon in the project Power Pylons of the Future (PoPyFu), the corona inception criterion applied in the present chapter is utilized. With the initial design of the fully composite pylon, corona discharge will be triggered from the composite cross-arm surface. Thus, steps must be taken to control the electric field magnitude on the cross-arm surface to an acceptable level. The application of corona rings is necessary.

5.5.2 Effects of Inclined Angles on Water Induced Corona Activities

As presented in Sect. 5.4.3, a higher inclined angle $\boldsymbol{\theta}_{cross-arm}$ leads to a lower corona inception voltage on the composite cross-arm surface in wet conditions. This trend may be explained by effects of the water droplet's shape on the electric field enhancement factor when the water droplet is presented on a polymer surface.

As discussed in Sect. 5.1.4, the background electric field is intensified at the triple junction where the polymer, air and water meet, as Fig. 5.6 shows. Several researchers have investigated effects of water droplets' characteristics on the electric field distribution on a polymer surface [13, 21–23]. It has been proven that when the water volume is the same, the electric field enhancement factor is bigger at the edge of an ellipsoidal water droplet compared with that of a spherical one. For example, the author found out that the background electric field was as low as 4 kV/cm on the surface of a silicone rubber without water droplets presented [21]. When a hemispherical water droplet with a diameter of 5 mm was presented on the silicone rubber surface, the maximum electric field was increased to 16.02 kV/cm. Further more, if the hemispherical water droplet was deformed into an ellipsoidal shape, with which the contact length between the water droplet and the polymer surface doubled, the maximum electric field was increased to 19.97 kV/cm. In this case, the electric field enhancement factor with a hemispherical water droplet was 4 approximately, being 5 approximately with an ellipsoidal water droplet. With a bigger electric field enhancement factor, the partial discharge (PD) inception voltage is reduced for the polymer surface in wet conditions.

Figure 5.29 shows different water droplet shapes with different cross-arm inclined angles in the water induced corona discharge test. From the figure, it can be seen that with increase of the inclined angle, the shape of water droplets on the weather sheath was transforming from hemispherical to ellipsoidal gradually (shown in blue frames). Thus, it is rational to expect a higher electric field with a higher cross-arm inclined angle, with the same rain density. As a result, the corona inception voltage is lower with a bigger cross-arm inclined angle.

Meanwhile, as Fig. 5.29 shows, bigger water droplets can be observed near the root of the shed with the increase of inclined angle, especially in the case of 30° (shown in the red frame in Fig. 5.29). It's explained in [13] that with increase of water droplet's volume, the electric field enhancement factor is increased. Thus, corona is more easier to happen with a bigger inclined angle, where the possibility of bigger water droplet's formation is higher compared with a smaller inclined angle.

It is also mentioned in [24] that the inclined angle of a polymer surface has effects on contact angles of water droplets deposited on it. Figure 5.30 indicates the definition of contact angles of a water droplet on a surface which is not horizontal. $\boldsymbol{\theta}_a$ and θ_r are defined as advancing angle and receding angle, respectively. α is the inclined angle of the surface. The value of θ_r is the most important when the hydrophobicity of a polymer surface shall be evaluated [24]. Table 5.2 gives the hydrophobicity classification proposed by STRI [24] based on evaluation of θ_r.

A smaller θ_r indicates a lower hydrophobicity level, as described in Table 5.2.

In the case of the composite cross-arm in the fully composite pylon, with an increasing inclined angle, θ_r of water droplets on the cross-arm surface is decreasing, until the minimum θ_r is reached and the water droplet will advance with a jerk [24]. Thus, with the same polymer material, an increasing inclined angle equals to decreasing the value of θ_r and thus decreasing the hydrophobicity of the material.

Test results presented in Sect. 5.4.3 also indicates that a higher inclined angle may lead to a less severe corona discharge from the composite cross-arm surface. For the

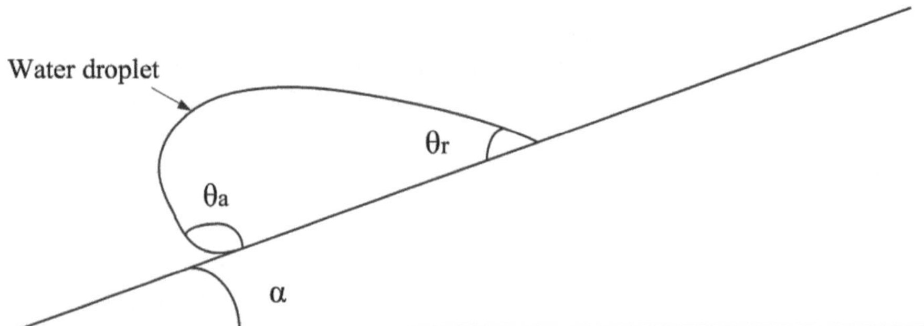

Fig. 5.30 Contact angles of a water droplet on a titled surface [24]

Table 5.2 Criteria for the hydrophobicity classification (HC) proposed by STRI [24]

HC	Description
1	Only discrete droplets are formed $\theta \approx 80°$ or larger for the majority of droplets
2	Only discrete droplets are formed $50° < \theta < 80°$ for the majority of droplets
3	Only discrete droplets are formed $20° < \theta < 50°$ for the majority of droplets Usually they are no longer circular
4	Both discrete droplets and wetted traces from the water runnels are observed (i.e. $\theta = 0°$). Completed wetted areas <2 cm^2. Together they cover $<90\%$ of the test area
5	Some completely wetted areas >2 cm^2, which cover $<90\%$ of the tested area
6	Wetted areas cover $>90\%$, i.e. small unwetted areas (spots/traces) are still observed
7	Continuous water film over the whole tested area

time being, the reason for this phenomenon is still unclear. More experimental data is needed for further analysis. Also, image processing technique may be needed for quantitative analysis of water droplets on the composite cross-arm surface.

5.6 Summary

In this chapter, electric field distribution on the composite cross-arm surface of the fully composite pylon has been evaluated by experimental method. Based on electric field distribution analysis, advices are given for the improvement of the cross-arm electrical design.

Electric field computation was performed in advance in [17] based on the initial design of the fully composite pylon and the electric field magnitude at 0.5 mm from the weather sheath of the cross-arm in dry conditions is 0.283 kV/mm, lower than the maximum acceptable value of 0.45 kV_{rms}/mm. Thus, corona discharge is not expected to happen on the composite cross-arm surface in wet conditions.

In order to verify the electric field distribution by experimental method, a high voltage test on a full-scale composite cross-arm segment was performed. The cross-arm segment was a conventional high voltage insulating bushing from ABB, Sweden, whose structure and configurations are almost the same with the initial design of the composite cross-arm in the fully composite pylon. Meanwhile, the electric field distribution on the insulating bushing surface with all test setups in place was computed and the maximum value of the electric field at 0.5 mm from the weather sheath of the bushing was consistent with that of the composite cross-arm. Thus, it is believed that test on the insulating bushing can be equivalent to test on the composite cross-arm, which hasn't been manufactured for the time being. The test results is a valuable reference for the design research of the composite cross-arm in the fully composite pylon.

Artificial rain was applied to the full-scale cross-arm segment with a 420 kV_{rms} phase-to-ground voltage applied. Corona discharge was measured from the weather sheath surface in wet conditions, which was absent in dry conditions. The water induced corona discharge inception voltage was recorded as 379 kV_{rms}, with which the maximum electric field magnitude at 0.5 mm from the weather sheath was calculated as 0.256 kV/mm. The corona inception electric field in wet conditions is lower than the maximum electric field magnitude on the composite cross-arm surface with nominal operation voltage, which indicates that water induced corona discharge will happen on the cross-surface in nominal operation. Thus, the configuration design of the cross-arm needs further consideration, aiming to restrain the maximum electric field magnitude at 0.5 mm from the weather sheath lower than 0.256 kV/mm.

Effects of the inclined angle of the cross-arm was also investigated in the chapter. The water induced corona discharge test was repeated with different cross-arm inclined angles and partial discharge (PD) characters were measured. The experimental results indicate that, due to varied water droplets morphology on the cross-arm surface, a bigger inclined angle leads to a lower corona inception voltage but a lower values of partial discharge (PD) peak, partial discharge (PD) average levels and partial discharge (PD) repetitive rate. Thus, the decision of the optimal inclined angle of the cross-arm in the fully composite pylon depends on the trade-off between corona inception voltage and corona discharge levels.

References

1. A.J. Phillips, J. Kuffel, A. Baker, J. Burnham, Electric fields on AC composite transmission line insulators. IEEE Trans. Power Deliv. **23**(2), 823–830 (2008)
2. N.G. Trinh, Partial discharge XIX: discharge in air part I: physical mechanisms. IEEE Electr. Insul. Mag. **11**(2), 23–29 (1995)
3. R. Hackam, Outdoor HV composite polymeric insulators. IEEE Trans. Dielectr. Electr. Insul. **6**(5), 557–585 (1999)
4. W. Song, W. Shen, G. Zhang, B. Song, Aging characterization of high temperature vulcanized silicone rubber housing material used for outdoor insulation. IEEE Trans. Dielectr. Electr. Insul. **22**(2), 961–969 (2015)
5. B. Vanica, T.K. Saha, T. Gillespie, *Australian Universities Power Engineering Conference*, Electric field modeling of non-ceremic high voltage insulator (Hobart, Australia, 2005)
6. A.J. Phillips, A.J. Maxwell, C.S. Engelbrecht, I. Gutman, Electric-field limits for the design of grading rings for composite line insulators. IEEE Trans. Power Deliv. **30**(3), 1110–1118 (2015)
7. P. Sidenvall, I. Gutman, L. Carlshem, J. Bartsch, Development of the water drop-induced corona test method for composite insulators. IEEE Electr. Insul. Mag. **31**(6), 43–51 (2015)
8. E. Kuffel, W.S. Zaengl, J. Kuffel, Electrostatic fields and field stress control, in *High Voltage Engineering - Fundamentals*, 2nd edn. (Butterworth-Heinemann, 2000), p. 203
9. Z. Guan, L. Wang, B. Yang, X. Liang, Electric field analysis of water drop corona. IEEE Trans. Power Deliv. **20**(2), 964–969 (2005)

10. I.J.S. Lopes, S.H. Jayaram, E.A. Cherney, A study of partial discharges from water droplets on a silicone rubber insulating surface. IEEE Trans. Dielectr. Electr. Insul. **8**(2), 262–268 (2001)
11. Y. Liu, B. Du, Pattern identification of surface flashover induced by discrete water droplets on polymer insulator. IEEE Trans. Dielectr. Electr. Insul. **21**(4), 1972–1981 (2014)
12. A.L. Souza, I.J.S. Lopes, Experimental investigation of corona onset in contaminated polymer surfaces. IEEE Trans. Dielectr. Electr. Insul. **22**(2), 1321–1331 (2015)
13. I.J.S. Lopes, S.H. Jayaram, E.A. Cherney, A study of surface discharges from water droplets on silicone rubber insulators, in *Conference on Electrical Insulation and Dielectric Phenomena: Annual Report* (Victoria, Canada, 2000), pp. 199–202
14. I. S. 60815, *Selection and dimensioning of high-voltage insulators intended for use in polluted conditions*, Std. 60 815 (2008)
15. ITEM24. Technical document of profile 8. https://product.item24.de
16. E.P. Popov, *Mechanics of Materials-SI Version*, 2nd edn. (Prentice/Hall International. Inc, 1978)
17. T. Jahangiri, Q. Wang, C.L. Bak, F.F. Silva, Electric stress computations for designing a novel unibody composite cross-arm using finite element method. IEEE Trans. Dielectr. Electr. Insul. **24**(6), 3567–3577 (2017)
18. M.D. Pfeiffer, Ion-flow environment of HVDC and hybrid AC/SC overhead lines, Ph.D. dissertation (ETH Zurich, 2017)
19. I. S. 60270, *High voltage test techniques - Partial discharge measurements*, Std. (2000)
20. C. Zachariades, S.M. Rowland, I. Cotton, V. Peesapati, D. Chambers, Development of electric-field stress control devices for a 132 kV insulating cross-arm using finite-element analysis. IEEE Trans. Power Deliv. **31**(5), 2105–2113 (2016)
21. Y. Zhu, S. Yamashita, N. Anami, M. Otsubo, C. Honda, Y. Hashimoto, Corona discharge phenomenon and behavior of water droplets on the surface of polymer in the AC electric field, in *7th International Conference on Properties and Applications of Dielectric Materials*, vol. 2 (Nagoya, Japan, 2003) pp. 638–641
22. N. Arise, H. Nishioka, Y. Okraku-Yienkyi, M. Otsubo, C. Honda, *Behavior of water droplet on polymer material and the influence on discharge characteristics*, vol. 2, Conference on Electrical Insulation and Dielectric Phenomena: Annual Report (USA, Austin, 1999), pp. 455–458
23. A.J. Phillips, D.J. Childs, H.M. Schneider, Water drop corona effects on full-scale 500 kV non-ceramic insulators. IEEE Trans. Power Deliv. **14**(1), 258–265 (1999)
24. S. G. 92/1, Hydrodrophobicity classification guide. Technical report, Tech. Rep

Chapter 6
Lightning Shielding Performance of Fully Composite Pylon

6.1 Introduction

One of the key issues in the design of any overhead line pylon is the evaluation of its lightning shielding performance. There are two typical methods to assess the lightning shielding performance of power pylons: the electro-geometric model (EGM) and the leader progression model (LPM) [1]. Traditionally, the lightning shielding performance of pylons is estimated on the basis of electro-geometric model (EGM). The EGM method is a design approach, which is dependent on the geometry of pylon and utilizes some empirical data regarding the observations of lightning incidences. This method gives an estimation for the expected distribution of lightning strokes on an overhead line, as well as an estimation for the expected number of the line outage caused by lightning [2]. The LPM method covers the physical process of lightning incidence and it is a more robust model to evaluate the lightning shielding performances, but it is not a widely used method at the engineering level due to its complexity, and thus, the EGM is still the most frequently used method for evaluating lightning shielding performance [2]. The guidelines for using EGM method are presented in IEEE Standard 1243:1997 [3] and CIGRÉ Working Group 33-01:1991 [4] and C4-26: 2017 [5]. CIGRÉ refers also to the LPM method. In this chapter, lightning shielding performance of fully composite pylon is evaluated based on a modified EGM method and the feasibility of assigning unusual shielding angle of $-60°$ for the pylon is investigated.

6.2 Shielding Angle

The shielding angle of high voltage power pylons depends on the position of shield wire(s) with respect to phase conductors. The shielding angles can be divided into two categories including positive and negative shielding angles as shown in Fig. 6.1.

T. Jahangiri et al., *Electrical Design of a 400 kV Composite Tower*, Lecture Notes in Electrical Engineering 557,
https://doi.org/10.1007/978-3-030-17843-7_6

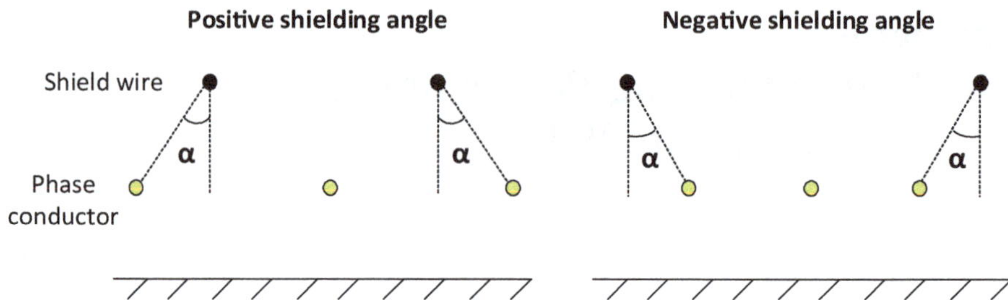

Fig. 6.1 Definition of shielding angle

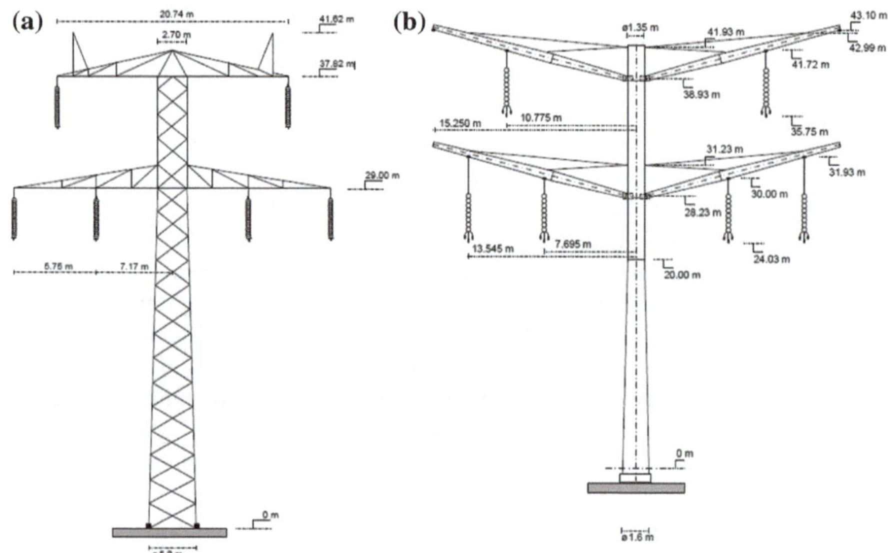

Fig. 6.2 **a** Donau pylon with positive shielding angle, **b** Eagle pylon with negative shielding angle [6]

Figure 6.2 also illustrates two pylons of "Donau" and "Eagle" which have positive shielding angle of 14.2° and negative shielding angle of −5.11°, respectively [6]. The effective shielding angle for an overhead line's pylon depends on the importance of the line and the value of Shielding Failure Flashover Rate (SFFOR), which can be determined by implementing EGM method.

For an overhead line design with critical load, a suitable value for shielding failure flashover rate (SFFOR) may be chosen as 0.05 outages/100 km-years (0.05 flashover/100 km-years) [3, 7] whereas the recommended value of SFFOR for general practice is between 0.1 and 0.2 outages/100 km-years [3, 8]. The predefined SFFOR can be achieved by adjusting the location of shield wire(s) at the tower top in a way that a specific number of outages occur during the Mean Time Between a shielding Failure flashover (MTBF). The relation between SFFOR and MTBF is as follows [9]:

$$MTBF = \frac{1}{L(SFFOR)} \quad \text{(year)} \tag{6.1}$$

L: is the length of line

MTBF for an overhead line with 100 km length and predefined shielding failure flashover rate (SFFOR) of 0.05 outages/100 km-years is:

$$MTBF = \frac{1}{L(SFFOR)} = \frac{100}{100(0.05)} = 20 \quad \text{(year)}$$

which means that the possible outage due to the incidence of lightning flashes to the phase conductors of the line is restricted to one during a period of 20 years.

In order to avoid outage, a shielding analysis should be done to derive perfect shielding angle for a pylon. A perfect shielding angle is an angle that provides zero SFFOR for the pylon [9, 10]. The perfect shielding angle can be achieved when lightning strokes with peak currents higher than the minimum current causing flashover of insulator (critical current) are intercepted [11]. The lightning stroke currents lower than critical current may not be prevented by shield wire(s) and attach to the phase conductors. Hence, these currents are not expected to cause outage due to flashover on phase insulation. In the case of incidence of lightning strokes to the overhead line's phase conductors close to substation, higher lightning stroke currents can cause a big issue for the substation. This can be resolved by utilizing surge arresters in the substation. Determination of effective and perfect shielding angles drastically depend on the implemented EGM models [11]. There are several EGM models, which distinguish due to differences in the calculation of striking distances. The striking distances of shield wires and phase conductors, r_c, are as a function of prospective lightning stroke current and the striking distance of earth surface, r_g, is proportional to r_c as follows:

$$r_c = AI^b = \gamma \cdot r_g \quad \text{(m)} \tag{6.2}$$

where:

r_c: Striking distance of shielding wire/phase conductor (m).
r_g: Striking distance of earth surface, $r_g = \frac{r_c}{\gamma}$ (m).
A, b and γ: Constant values based on different EGM models which are given in Table 6.1.
I: Lightning current (kA).

Since striking distances differ depending on the constant values of A, b and γ, therefore, utilizing different EGM models leads to obtaining different values for the effective and perfect shielding angles [11]. However, the acceptable shielding angles for overhead lines with different voltage levels and heights are reported in

Table 6.1 Evaluation of striking distances using different sources [6]

EGM methods			
Source	A	b	γ
Young	27γ	0.32	γ_y
Brown-Whitehead	7.1	0.75	1.11
Love	10	0.65	1
IEEE-1991 T&D Committee	8	0.65	$1/\beta^a$
IEEE Std. 1243 – 1997	10	0.65	$1/\beta^b$
Wagner & Hileman	14.2	0.42	1
Mousa & IEEE – 1995 Substation Committee	8	0.65	1

where: $\beta^a = 22/y$, $0.6 < \beta < 0.9$, y is the phase conductor height. $\beta^b = 0.36 + 0.17 \ln(43 - y_c)$ if $y_c < 40$ and $\beta^b = 0.55$ if $y_c \geq 40$, y_c is the phase conductor height. $\gamma_y = 1$ for $h < 18$ m and $\gamma_y = 444/(462 - h)$ for $h > 18$ m, h is the ground wire height

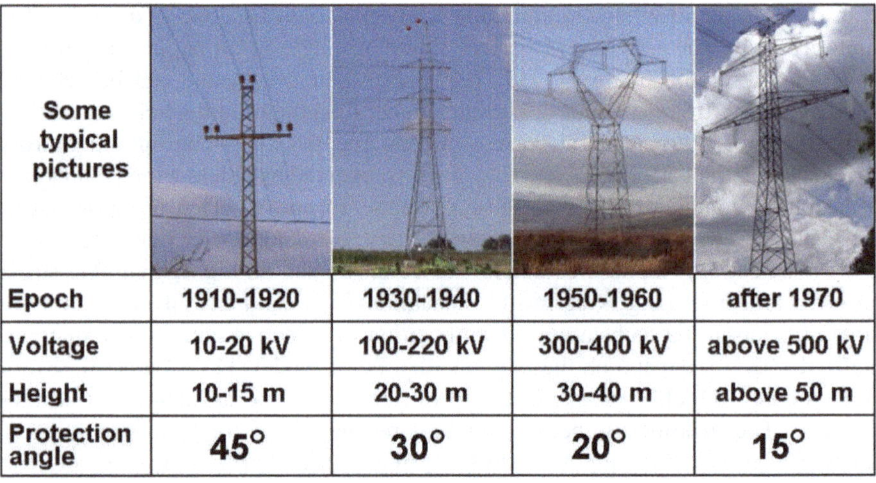

Fig. 6.3 Shielding angle based on SFFOR of 0.1 outages/100 km-years [12]

[12] which are depicted in Fig. 6.3. These shielding angles are collected based on overhead transmission line's operation experiences over 100 years.

From Fig. 6.3, it can be seen that the required shielding angle decreases by increasing voltage levels and height of pylons. The difference in the shielding angles of overhead lines is due to their different tower structure shapes, tower heights, voltage level of lines, weather conditions and the importance of overhead lines in power systems. The preliminarily assigned shielding angle of −60° with a height of 22.5 m are unique features of fully composite pylon. Therefore, the predefined shielding angle and the proper shielding angle for the fully composite pylon should be derived from lightning shielding analyses, which are presented in next sections.

6.3 Shielding Analysis Using Electro-Geometric Method (EGM)

The EGM model based on IEEE Std. 1243–1997 [3] is representative for the application of EGM method. The general concept of EGM method for conventional pylons is presented in [9] and depicted in Fig. 6.4 in which the arcs (striking distances) with the radius of r_c are drawn from shield wires and phase conductors; the horizontal line above the ground surface is constructed for the striking distance of ground surface r_g and the striking distances of r_c and r_g are derived from a specific lightning stroke current. The intersections between arcs and the intersection between arcs and horizontal line are pointed by A, B and C. According to EGM theory, the incidence of direct lightning strokes to the region between A and B ($\overset{\frown}{AB}$) terminates to the phase conductor. The incidence to the region between B and C ($\overset{\frown}{BC}$) d wires and the incidence beyond A terminates to the ground surface. From Fig. 6.4a, it can be seen that in a traditional overhead line tower, the region around point C is properly protected by overlapping two striking distances of shield wires. However, there are two regions in both sides of the overhead line's pylon that are not protected against vertical lightning strokes and are defined by the distance of D_c in Fig. 6.4a. The distances of D_g and S_g are also the exposure distances for the shield wires.

Figure 6.4b shows that the striking distances increase by increasing the amplitude of lightning stroke current. There is a specific current that causes the unprotected distance of D_c becomes zero and all striking distances of shield wire, phase conductor and ground meet at a single point. This current is called the Maximum Shielding Failure Current (I_{MSF}) and it is the current at and above which lightning strokes will not terminate to the phase conductor [9]. As a result, the phase conductors of

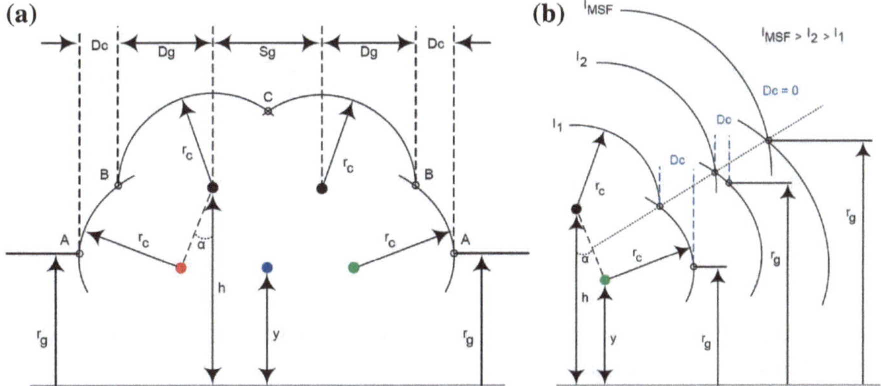

Fig. 6.4 **a** Definitions of angles and striking distances, **b** Definition of maximum shielding failure current (I_{MSF}) where $D_c = 0$ [6]

overhead line are properly protected against lightning stroke for currents higher than I_{MSF}.

The Maximum Shielding Failure Current (I_{MSF}) can be calculated by:

$$I_{MSF} = \left(\frac{\gamma \cdot \frac{h+y}{2}}{A(1 - \gamma \cdot \sin\alpha)} \right)^{\frac{1}{b}} \quad \text{(kA)} \tag{6.3}$$

where, h and y are the height of shield wire and phase conductor, respectively. A, b and γ are given in Table 6.1, and α is the shielding angle.

Maximum Shielding Failure Current (I_{MSF}) is used to determine shielding failure rate (SFR) and shielding failure flashover rate (SFFOR) of overhead lines, which are criteria for the acceptability of lightning performance of the lines. Shielding failure rate (SFR), is the annual number of lightning strikes to phase conductors of an overhead line with the length of 100 km, which lead to shielding failure. A shielding failure with a low lightning stroke current may not necessarily cause a flashover and therefore outage of the line. Shielding failure rate (SFR) can be calculated by [3, 5]:

$$SFR = 0.2N_g \int_{\mathrm{I_{min}}}^{\mathrm{I_{max}}} D_c(I) \cdot f(I)_1 dI \quad \text{(Flashes/100 km-year)} \tag{6.4}$$

- I_{max}: Upper limit of integration which is equal to Maximum Shielding Failure Current (I_{MSF})
- I_{min}: Lower limit of integration which is zero based on [3, 5] and 3 kA based on [4]
- N_g: Ground flash density (for Denmark is 1.39 Flashes/km^2-year).
- $D_c(I)$: Exposure width of phase conductor which is as a function of stroke current and given by Eq. (6.5) and Fig. 6.5 [6]

$$D_c(I) = r_c\left[\cos\left(\sin^{-1}\left(\frac{r_g - y}{r_c}\right)\right) - \cos\left(\tan^{-1}\left(\frac{a}{h-y}\right) + \sin^{-1}\left(\frac{c}{2r_c}\right)\right)\right] \quad \text{(m)} \tag{6.5}$$

$f(I)_1$: Probability density of first stroke current I and is given by:

$$f(I)_1 = \frac{1}{\sqrt{2\pi} \cdot \beta \cdot I} \cdot e^{-\frac{1}{2}z^2} \quad , \quad z = \frac{\ln\left(\frac{I}{M}\right)}{\beta} \tag{6.6}$$

where, β and M are logarithmic standard deviation and median value of striking current I, respectively and given by:

$$I \le 20\text{kA}\ \text{M} = 61.1\ \text{kA}\ \beta = 1.33$$
$$I > 20\text{kA}\ \text{M} = 33.3\ \text{kA}\ \beta = 0.605$$

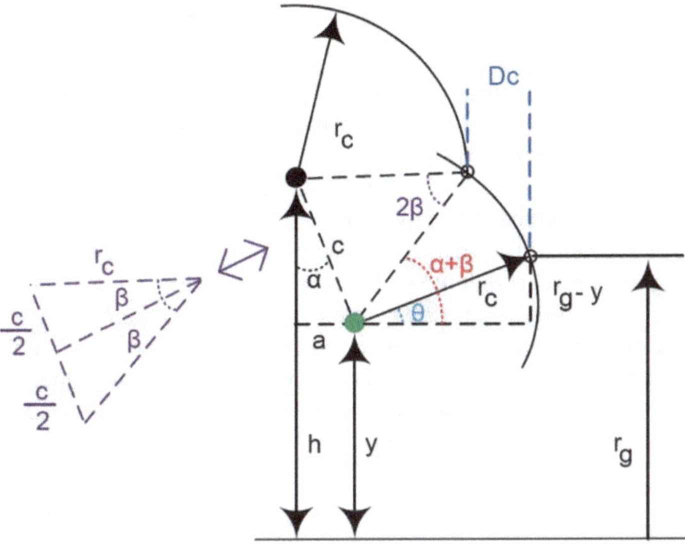

Fig. 6.5 Illustration of parameters to calculate $D_c(I)$ [6]

Shielding failure flashover rate (SFFOR), is the annual number of lightning strikes to phase conductors which lead to flashover of line insulation (and subsequently result in outage of the line) and can be expressed by [3, 5]:

$$SFFOR = 0.2N_g \int_{I_{min}=I_C}^{I_{max}=I_{MSF}} D_c(I) \cdot f(I)_1 dI \quad \text{(Outages/100 km-year)} \tag{6.7}$$

where, I_c is the minimum shielding failure current which can cause flashover of line insulation. The critical current (I_C) (the minimum shielding failure flashover current) can be calculated by [3, 13]:

$$I_C = \frac{2CFO}{Z_S} \quad \text{(A)} \tag{6.8}$$

where, CFO is the negative polarity critical flashover voltage of line insulator.
Z_S is the conductor surge impedance under corona

According to Eq. (6.7), flashover on line insulation will not occur if the perspective lightning stroke current is below the critical current. On the other hand, lightning flashes may have subsequent lightning stroke currents higher than first stroke current and may produce crest currents greater than the critical current and therefore they could be responsible for the outage of overhead lines. In this study, the effectivity of shielding angles are evaluated for the first stroke current and the effect of subsequent lightning stroke currents is ignored in the lightning shielding analysis of fully composite pylon.

6.4 Fully Composite Pylon with −60° Shielding Angle

The basic design of fully composite pylon is based on a 30° inclination angle, which is chosen to improve the visual appearance of pylon. Hence, the lightning shielding performance of fully composite pylon with −60° shielding angle is evaluated in this section. Inclination angle below 40° increases the separation distance between two shield wires. An overhead line pylon should protect its phase conductors against lightning stroke currents at and above the Maximum Shielding Failure Current (I_{MSF}). However, Fig. 6.6a reveals that the inclination angle of 30° imposes an unprotected zone at the center of fully composite pylon by considering a lightning stroke current equal to Maximum Shielding Failure Current (I_{MSF}). It can be seen that the striking distances of phase conductors, which are drawn for Maximum Shielding Failure Current (I_{MSF}), are in the exposure of vertical lightning strokes at the center of fully composite pylon (in the region between C, D and C′) whereas the lateral sides of the pylon ($\widehat{BC}$ and $\widehat{B'C'}$) are fully protected.

The unprotected zone at the center of pylon is illustrated in Fig. 6.6b, which is specified by a red area. In order to evaluate the lightning shielding performance of fully composite pylon with 30° inclination angle, the probability of incidence of lightning strokes to the unprotected distance of U_c should be estimated. The unprotected distance of U_c is as a function of lightning stroke current and can be given by Eq. (6.9), which is determined based on Fig. 6.7:

$$U_c(I) = S_g - 2S_T, \quad S_T = r_c \sin\left(\theta_{arm} + \frac{\delta}{2}\right) \Rightarrow U_c(I) = S_g - 2r_c \sin\left(\theta_{arm} + \frac{\delta}{2}\right) \tag{6.9}$$

In Fig. 6.7, D_{ph_shw} and D_{ph_ph} are the air clearances on the fully composite pylon. h is the height of shield wires and S_g is the separation between two shield wires. θ_{arm} is also the inclination angle of the unibody cross-arm. The radius of circles, r_c, in Fig. 6.7a, is the striking distance of shield wires/phase conductors and the height of horizontal line, r_g, is the striking distance of earth surface.

The value of r_c is somewhat higher than r_g, which is due to this fact that the electric field gradients around shield wires/phase conductors are somewhat greater than at the earth surface [14]. The striking distances shown in Fig. 6.7a are related to a lightning stroke current equal to Maximum Shielding Failure Current (I_{MSF}).

The unprotected distance of $U_c(I)$ increases by decreasing striking distance and corresponding lightning stroke current, and in the same way, decreases by increasing striking distance and the relevant lightning stroke current. Meanwhile, there is a current at which the value of $U_c(I)$ becomes zero. This current is so-called Minimum Intersection Current (I_{MIC}) which corresponds to a situation that two shield wires fully protect the center of fully composite pylon with 30° inclination angle as shown in Fig. 6.8.

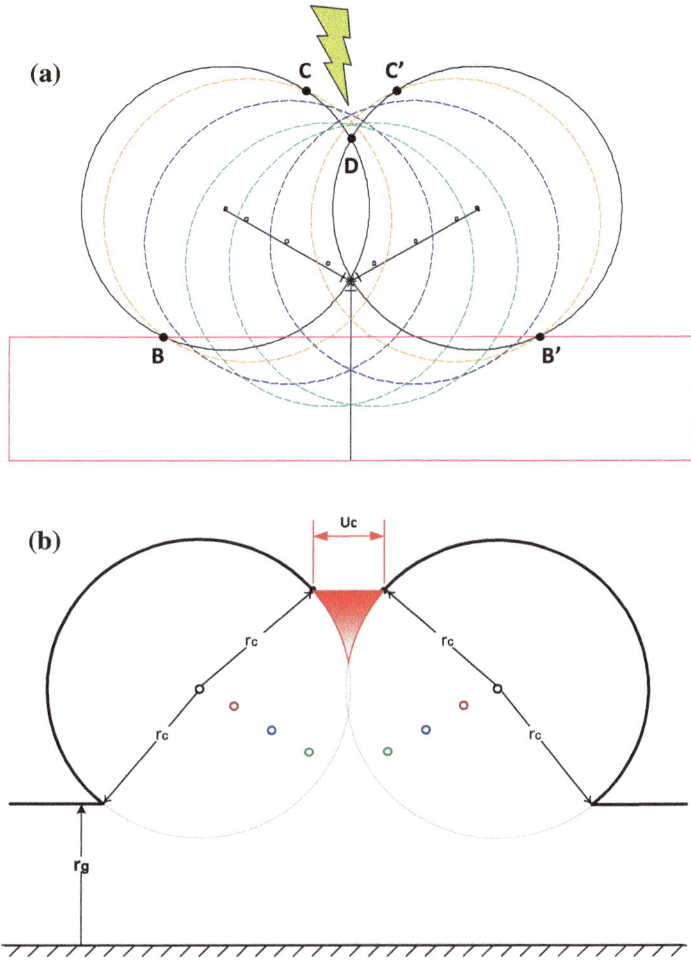

Fig. 6.6 **a** Unprotected zone at the center of fully composite pylon, **b** Definition of U_c

Figure 6.8 shows that the unprotected zone at the center of pylon disappears for the lightning stroke currents equal to I_{MIC}, which is remarked by point C. It is evident that the value of Minimum Intersection Current (I_{MIC}) is higher than Maximum Shielding Failure Current (I_{MSF}). I_{MIC} is the maximum lightning stroke current that can lead to shielding failure in fully composite pylon with 30° inclination angle. It should be mentioned that two points of B and B′ are shifted below and two new intersection points of E and E′ are pointed out in Fig. 6.8. Minimum Intersection Current (I_{MIC}) can be determined by solving the following equations:

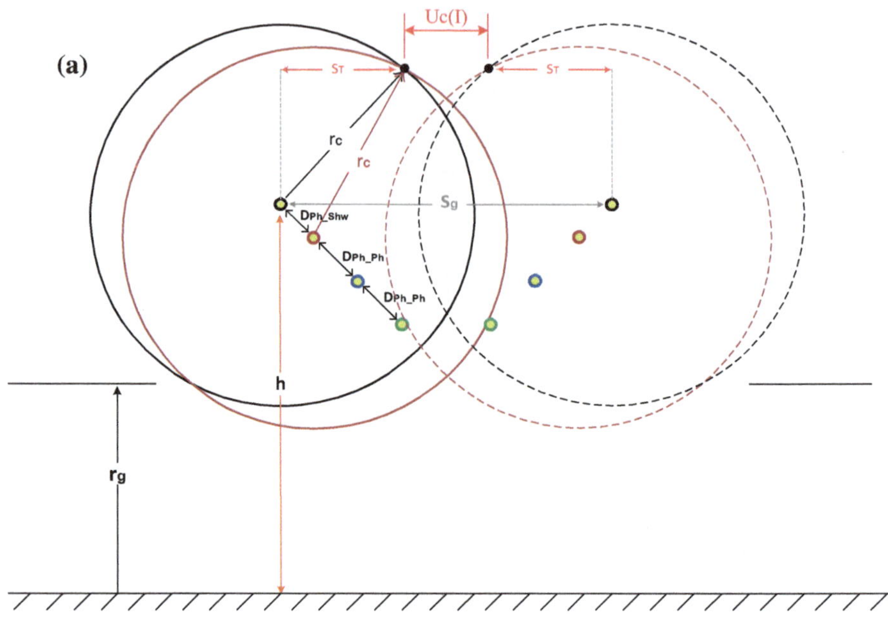

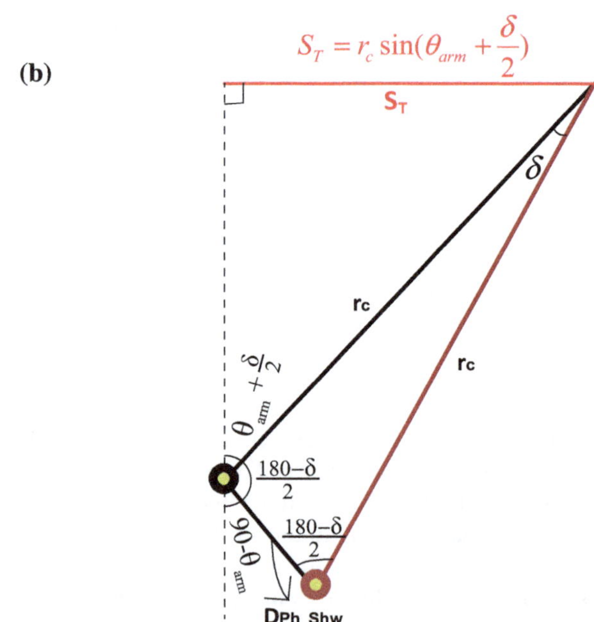

Fig. 6.7 **a** Determination of $U_c(I)$ based on pylon geometry and striking distances, **b** Definition of parameter S_T

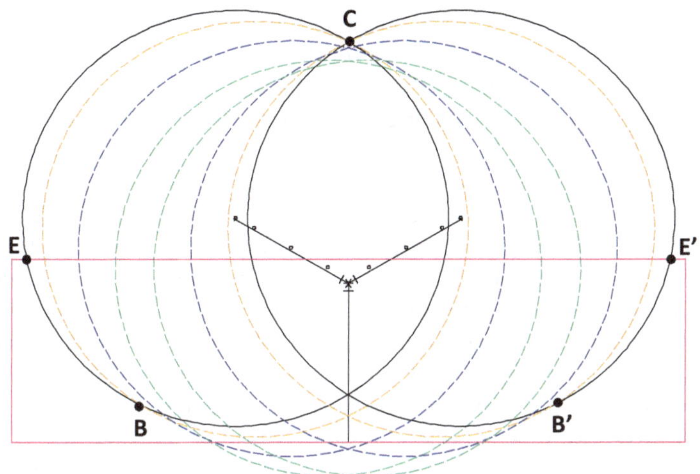

Fig. 6.8 Protection of fully composite pylon for the lightning stroke current of I_{MIC}

$$U_c(I) = 0 \quad \Rightarrow \quad S_g - 2r_c \sin\left(\theta_{arm} + \frac{\delta}{2}\right) = 0 \tag{6.10}$$

$$S_g - 2r_c \sin\left(\theta_{arm} + \frac{1}{2} Arc \cos\left(1 - \frac{(D_{Ph\_Shw})^2}{2(r_c)^2}\right)\right) = 0 \tag{6.11}$$

where, $r_c = AI^b$ and $\delta = \text{Arc}\cos\left(1 - \frac{(D_{Ph\_Shw})^2}{2r_c^2}\right)$. The lightning stroke current of I_{MIC} can be determined by interpolating Eq. (6.12) for a variable of I:

$$S_g - 2AI^b \sin\left(\theta_{arm} + \frac{1}{2} Arc \cos\left(1 - \frac{(D_{Ph\_Shw})^2}{2(AI^b)^2}\right)\right) = 0 \tag{6.12}$$

As it mentioned before, traditional overhead transmission pylons have two lateral unprotected zones at the both sides of the pylons whereas the unprotected zone on the fully composite pylon is at the center of the pylon. Since there is a difference between the unprotected zones of fully composite pylon (with −60° shielding angle) and other traditional pylons, the conventional EGM method cannot directly be used to calculate SFR and SFFOR of the fully composite pylon [14]. Introducing Minimum Intersection Current (I_{MIC}) instead of Maximum Shielding Failure Current (I_{MSF}) is the reason that a revised EGM method should be adopted to calculate SFR and SFFOR of the pylon with 30° inclination angle. In the modified EGM method, the unprotected zone of $U_c(I)$ in Eq. (6.9) (instead of the unprotected distances of $D_c(I)$ in Eq. (6.5)) is used in Eqs. (6.4) and (6.7) for the calculation of SFR and SFFOR of fully composite pylon.

The lightning shielding performance of fully composite pylon with −60° shielding angle is presented in [14], which has been evaluated based on the modified EGM and several attachment models given in Table 6.1. Based on different EGM models in [14], the calculated striking distances for the pylon differ slightly, but there is a variability between the calculated values of Maximum Shielding Failure Currents (I_{MSF}) and Minimum Intersection Currents (I_{MIC}). Moreover, the values of Maximum Shielding Failure Currents (I_{MSF}) are much lower than that for conventional pylons and are below 3 kA. The calculated values of Minimum Intersection Currents (I_{MIC}) are between 2.5 kA and 4.5 kA except in one case that is about 0.5 kA. The lower values of Minimum Intersection Currents (I_{MIC}) implies that lightning flashes with low intense can hit the phase conductors of the pylon. Since the critical current of pylon's insulation is much higher than the Minimum Intersection Currents (I_{MIC}), therefore, the line insulation can withstand against induced overvoltages caused by lightning flashes with low crest currents. Furthermore, shielding failure flashover rate (SFFOR) of fully composite pylon becomes zero because of higher critical current and theoretically, occurrence of line's outage seems to be unlikely.

The striking distances, I_{MSF} and I_{MIC} of fully composite pylon with $-60°$ shielding angle are given in Table 6.2. In order to better understanding the lightning performance of fully composite pylon with $-60°$ shielding angle, its lightning protection level is compared with the two pylons of Eagle and Donau towers, which are shown in Fig. 6.2. Striking distances of shield wires/phase conductors and earth surface together with Maximum Shielding Failure Currents (I_{MSF}) and critical currents (I_c) are given in Table 6.2 for the three pylons. The Eagle and Donau pylons provide full protection for phase conductors against lightning stroke currents higher than their Maximum Shielding Failure Currents, whereas fully composite pylon with $-60°$ shielding angle provides full protection for the lightning stroke currents higher than Minimum Intersection Currents (I_{MIC}).

Table 6.2 shows that striking distances of Donau pylon are about twice the striking distances of Eagle pylon and the striking distances of fully composite pylon are approximately 2.5 times lower than the striking distances of Eagle pylon. The reason for variability of striking distances between three pylons is due to this fact that striking distance is related to Maximum Shielding Failure Current (I_{MSF}) and it is a nonlinear function of shielding angle and pylon's height. Therefore, the drastic reduction in the striking distances of fully composite pylon (in comparison to two other pylons) is mainly due to its compact design, lower height and shielding angle. Table 6.2 also shows that the Maximum Shielding Failure Current of the fully composite pylon is also much lower than corresponding currents for the two other pylons. The critical currents given in Table 6.2 also differ because they depend on the line insulator's *CFO* and surge impedance of conductor under corona.

Shielding failure rate (SFR) and shielding failure flashover rate (SFFOR) of three pylons are given in Table 6.3. It can be seen that SFR of fully composite pylon is much lower than Eagle and Donau pylons and is almost zero. This implies that shield wires in both tips of fully composite pylon effectively protect phase conductors against lightning flashes. Moreover, it reveals that a pylon with negative shielding angles could have lower SFR in compassion to a pylon with positive shielding angle. SFFORs of fully composite pylon and Eagle pylon are zero whereas Donau pylon has a non-zero value for SFFOR. The SFFORs of two pylons are zero because their

Table 6.2 Evaluation of striking distances EGM model of IEEE 1243-1997 for different pylons

Electro-geometric model (EGM)	Fully composite pylon (−60° shielding angle)					Eagle pylon (−5.11° shielding angle)				Donau pylon (14.2° shielding angle)			
	r_c (m)	r_g (m)	I_C (kA	I_{MSF} (kA)	I_{MIC} (kA)	r_c (m)	r_g (m)	I_C (kA)	I_{MSF} (kA)	r_c (m)	r_g (m)	I_C (kA)	I_{MSF} (kA)
IEEE Std. 1243–1997	12.5	11.0	8.2	1.4	3.1	35.3	30.4	17.8	6.9	75.5	56.0	18.5	22.4

Table 6.3 SFR and SFFOR results for three pylons based on EGM model of IEEE 1243-1997

Type of pylon	SFR (Flashes/100 km-year)	SFFOR (Outages/100 km-year)
Fully composite pylon (−60° shielding angle)	0.0008	0
Eagle pylon (−5.11° shielding angle)	0.0353	0
Donau pylon (14.2° shielding angle)	0.1150	0.0234

critical currents (I_c) are higher than their corresponding Maximum Shielding Failure Currents (I_{MSF}), and therefore, according to Eq. (6.7), theoretically the possibility for the outage of both lines are zero but practical experiences show that there are still shielding failures on the pylons with zero SFFOR's.

Based on Table 6.2, the value of critical current for Donau pylon is lower than its Maximum Shielding Failure Current, which causes a non-zero value for SFFOR. The comparison of SFRs shows that fully composite pylon with −60° shielding angle has better lightning performance than other traditional pylons and the fully composite pylon with −60° shielding angle (30° inclination angle for the unibody cross-arm) provides satisfactory protection against lightning strikes. In order to further evaluate the lightning performance of fully composite pylon, the Rolling Sphere Method is also adopted in next section to verify the effectiveness of assigned shielding angle of −60° for the pylon.

6.5 Shielding Analysis Using Rolling Sphere Method (RSM)

For the purpose of power system substations, a well-known method to examine the effectiveness of assigned shielding system is utilizing rolling sphere method (RSM), which is based on electro-geometrical (EGM) theory. Rolling sphere method is a simplified method to apply EGM theory for the shielding of substations and is reported in IEEE Guide for Direct Lightning Stroke Shielding of Substations [13]. The rolling sphere procedure involves rolling of an imaginary sphere with a specific radius over and around of the shield wires to protect bus-bar [13]. Phase conductors of bus-bar are protected from direct lightning stroke if they remain below the surface of sphere without any contact or penetration into the sphere. The rolling sphere method is also introduced in [12, 15] as a method to evaluate the lightning shielding performance of high voltage overhead lines. Although, it cannot determine the efficiency of the lightning shielding system [12, 16].

Normally, overhead transmission lines have positive shielding angles, which can almost fully protect the center phases in comparison to outer phases. As it is mentioned before, fully composite pylon has a negative shielding angle of −60°. The

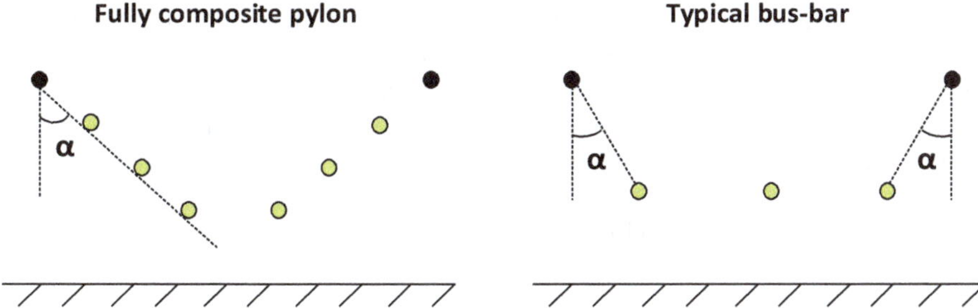

Fig. 6.9 Negative shielding angles of fully composite pylon and a typical bus-bar

similarity of having negative shielding angle between the fully composite pylon and a typical bus-bar with two shield wires, shown in Fig. 6.9, implies that the evaluation of lightning performance of fully composite pylon can be done using rolling sphere method and the results can have significant importance. Because the lightning protections of high voltage substations have higher importance in comparison of overhead lines and they are widely evaluated using rolling sphere method. Therefore, the application of rolling sphere method for an overhead line would also present a good estimation about the lightning performance of the overhead line.

Similar to the application of EGM method for overhead transmission lines, a specific SFFOR should be adopted for the lightning shielding of substations, but its application is difficult for the protection of equipment [9]. Therefore, for the simplicity, the design of protection system in substations is based on a design current. In order to evaluate lightning performance of a substation, determination of the design current (I_d) is necessary. The design current is as the same as critical current (I_c) in EGM method, but it is usually larger and can be calculated by [9]:

$$I_d = (1.27 + 0.72e^{-Ng/4})I_c \tag{6.13}$$

where:

N_g: Ground flash density
I_c: Critical Current (Minimum shielding failure flashover current)

By considering ground flash density of N_g (in Denmark) as 1.39 flashes/(km^2.year), and the critical current of 8.2 kA, the value of design current is 14.6 kA. This value is higher than the recommended value for design current for a system voltage above 230 kV which is 10 kA [9, 17]. On the other hand, it is indicated in IEEE Std. 998-2012 [13] that the protective zone of shield wires depends on the magnitude of lightning stroke current and if the shield wires protect phase conductors for the critical current of I_c, they may not shield the phase conductors for the lightning stroke currents lower than I_c. Conversely, any lightning stroke current higher than I_c can be better protected by shield wires because the higher currents will have greater striking distances. However, lightning stroke currents less than I_c can penetrate to the shielded system and terminate to the protected conductors, therefore, the insulation of system and equipment (their BIL levels) should be able to withstand against the induced voltages and prevent from the occurrence of flashover on the insulations. As a result, the minimum lightning stroke current that can lead to insulator flashover should be used as design current [13, 15] and therefore the critical current of 8.2 kA is chosen as the design current for the application of rolling sphere method for fully composite pylon.

6.5.1 Protected Areas and Striking Distances in Rolling Sphere Method

The EGM theory indicates that the protective area of shield wires depends on the amplitude of the stroke current and the rolling sphere radius can be determined based on the value of adopted design current. In conventional EGM method for the overhead transmission lines, the striking distances of shield wires and phase conductors are equal. However, in the EGM method of "IEEE-1995-IEEE Substations Committee", there are three striking distances which are equal, including:

- Striking distance to the ground surface, $r_g = 8I^{0.65}$, I lightning stroke current in kA
- Striking distance to the shield wire, $r_s = \gamma_s r_g, \quad (\gamma_s = 1)$
- Striking distance to the object (here phase conductors), $r_c = \gamma_c r_g, \quad (\gamma_c = 1)$

The protected zone provided by two shield wires is displayed in Fig. 6.10 in which three striking distances of r_g, r_s, and r_c are shown. Other parameters in Fig. 6.10 are as follows:

y_{mc}: Minimum protective height
h: Height of shield wires

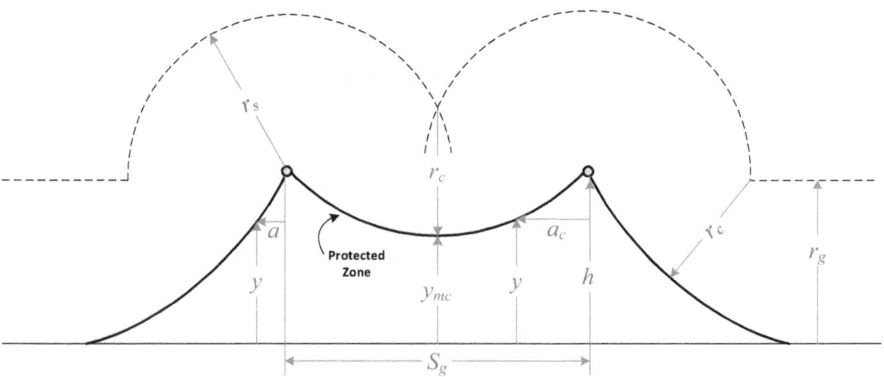

Fig. 6.10 Parameters of protected zone provided by two shield wires in RSM

Fig. 6.11 Determination of height of shield wires

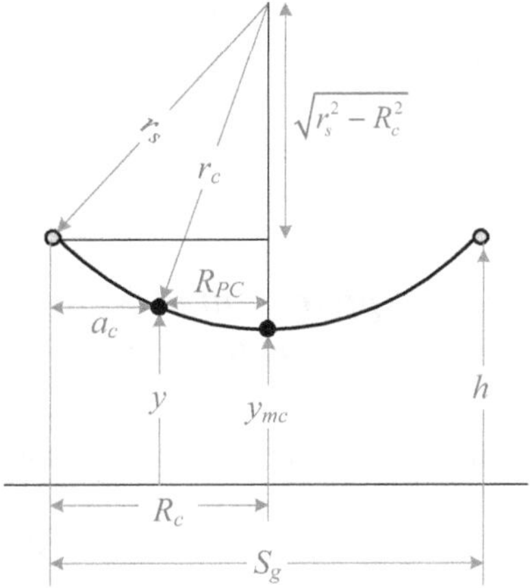

y: Height of the object to be protected (phase conductor)
S_g: Separation between shield wires
a_c: Horizontal distance between the object (phase conductor) and shield wire

In Fig. 6.10, two arcs from shield wires with the radius of r_s meet each other at a location, which is higher than the striking distance of ground surface, r_g. Therefore, any vertical stroke approaching the region between shield wires will terminate to the arcs of shield wires, r_s. Moreover, the striking distances of r_c, drawn from the intersection of arcs of shield wires (r_s) with the horizontal line of r_g, have provided two extra protected zones at the regions outside the shield wires, as shown in Fig. 6.10. According to Fig. 6.11, the protected zone between two shield wires can be explained by two distances of R_{pc} and a_c which are [9]:

$$R_{pc} = \sqrt{r_c^2 - \left(h - y + \sqrt{r_s^2 - R_c^2}\right)^2} \tag{6.14}$$

$$a_c = R_c - R_{pc} \tag{6.15}$$

where, $R_c = \frac{S_g}{2}$

The minimum protective height of y_{mc} between the shield wires can be calculated by setting $R_{pc} = 0$:

$$y_{mc} = h - r_c + \sqrt{r_s^2 - R_c^2} \tag{6.16}$$

By known the position of the object (phase conductors), Eq. (6.16) can be used to determine the required height for the shield wires:

$$h = y_{mc} + r_c - \sqrt{r_s^2 - R_c^2} \tag{6.17}$$

Two shield wires at both tips of the fully composite pylon have a specific height, therefore, the lightning performance of fully composite pylon based on pre-specified values for the position of phase conductors and shield wires should be evaluated for different lightning stroke currents.

6.5.2 *Application of Rolling Sphere Method for Fully Composite Pylon*

In this section, some different lightning stroke currents are chosen and the probability of their occurrence are presented. Anderson [13, 18] proposed a simple equation for the approximation of log-normal distribution for lightning stroke current as:

$$P(I) = \frac{1}{1 + \left(\frac{I}{31}\right)^{2.6}} \tag{6.18}$$

where,

$P(I)$: Probability that the peak current in any stroke will exceed I
I: Specified crest current of stroke in kA

The probabilities of occurrence of different lightning stroke currents given in Table 6.4 are estimated by using Andersons' cumulative distribution of lightning stroke current. Table 6.4 shows that the probability of appearance of lightning stroke currents below 3 kA is 0.003. For the lightning stroke currents less than 3 kA, the proposed value by CIGRE is 3 kA and this current has been widely used as the minimum value for lightning stroke current [9].

Furthermore, termination of lightning currents with this value to the phase conductors cannot lead to shielding failure because the insulation of system will withstand against the generated overvoltages. In the other word, striking lightning strokes to the phase conductors with the currents higher than the critical current ($I_c = 8.2$ kA) can cause shielding failure in system because the flashover on insulation will occur. However, in order to investigate the possibility of shielding failure on the fully composite pylon, different lightning stroke currents are examined and subsequently, the resultant protected zones are visually illustrated and discussed.

Table 6.4 Peak current distributions

	Peak current (I)				
	3 kA	5 kA	8.2 kA	10 kA	50 kA
Rolling sphere radius (R)	16.3 m	22.8 m	31.5 m	35.8 m	101.8 m
Probability of greater value	0.997	0.991	0.969	0.949	0.223

6.5.2.1 Lightning Stroke Current of 3 kA

The striking distances (with the radius of R) corresponding to the lightning stroke current of 3 kA are shown in Fig. 6.12. The blue dashed lines represent a sphere with the radii of R, which has intersections with two shield wires at both tips of the unibody cross-arm. By rolling the sphere on the top of the fully composite pylon, Fig. 6.12a shows that the top and middle phase conductors are placed inside the sphere. Conversely, the lower phase conductors are at the outside of the sphere. It means that the lower phase conductors are fully protected whereas the top and middle phase conductors are located at the outside of protected zone specified by red color, and therefore, are exposed to the lightning stroke currents equal to 3 kA and below. Comparing this current with the value of Minimum Intersection Current (I_{MIC} of 3.102 kA) in the modified EGM method shows that both methods predict a possibility of shielding failure for the fully composite pylon. In Fig. 6.12b, the objects (phase conductors) placed in red zone are protected against the lightning strokes. Although the phase conductors in the green zone are in the exposure of the lightning strokes, the insulation of unibody cross-arm can withstand against the produced lightning overvoltages and the possibility of occurrence of flashover on the insulation of unibody cross-arm is unlikely.

6.5.2.2 Lightning Stroke Current of 5 kA

The protected area against lightning stroke current of 5 kA is illustrated in Fig. 6.13. It can be seen that the middle and lower phase conductors are within the protected zone (red zone) but the top phase conductors are in the border between the red and green zones.

This current can be introduced as a threshold current for the lightning strokes and any current amplitude above this current will cover all phase conductors in the red zone (protected zone). Since this lightning stroke current is lower than the critical current ($I_c = 8.2$ kA), the insulation of unibody cross-arm can withstand against the overvoltages due to striking of lightning to the top phase conductors.

6.5.2.3 Design Current—Critical Lightning Current of 8.2 kA

The critical current of 8.2 kA is chosen as the design current for the protection of fully composite pylon. Termination of this lightning current to any phase conductor can produce an overvoltage which is equal to the critical flashover voltage for the insulation of unibody cross-arm and thus, can damage the cross-arm of fully composite pylon. However, Fig. 6.14 shows that all phase conductors are inside the protected zone (red zone) and the sphere on the fully composite pylon do not have any contact with the phase conductors. Therefore, it can be said that this lightning stroke current has been properly protected by shield wires and the lightning currents higher than the critical currents would be effectively attracted by shield wires. Higher lightning stroke currents have larger striking distances and therefore give higher radius for the

(a)

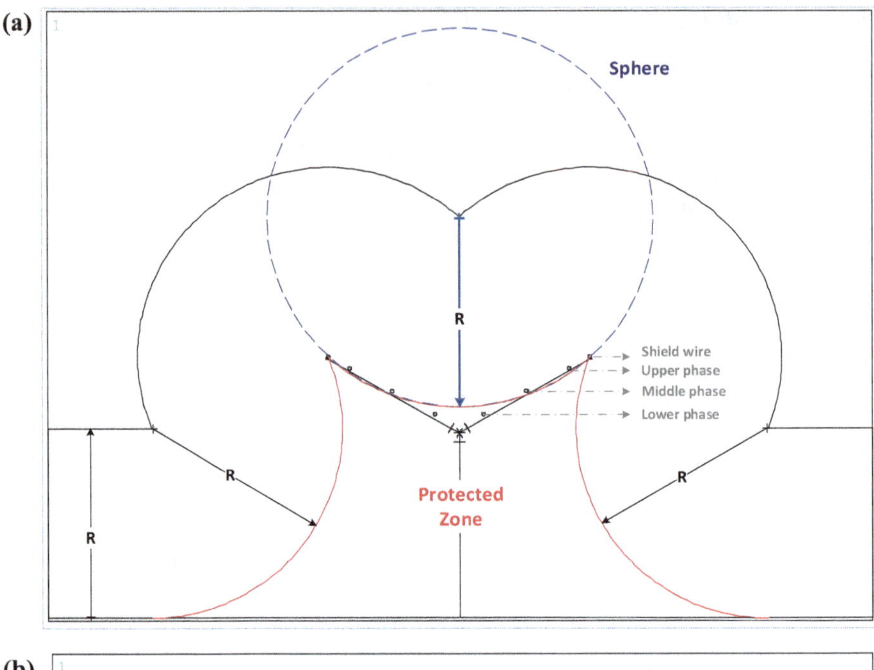

(b)

Fig. 6.12 **a** Striking distances with the radius of R = 16.3 m for I = 3 kA, **b** protected conductors (in red area) and unprotected conductors (in green area)

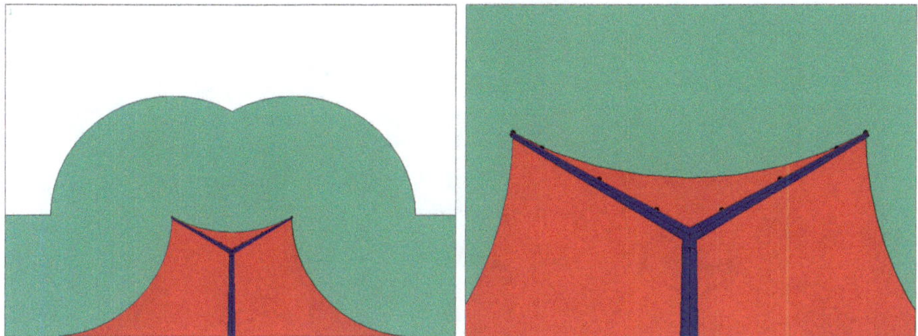

Fig. 6.13 Full view and zoomed view of protected conductors in red area and at the border of green area for lightning stroke current of 5 kA

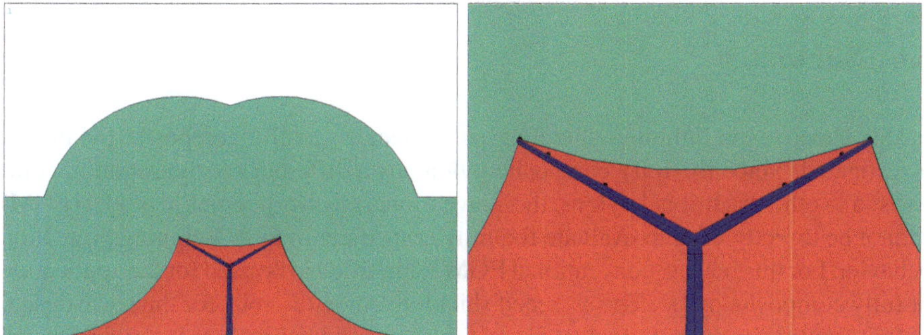

Fig. 6.14 Full view and zoomed view of protected conductors in red area for lightning stroke current of 8.2 kA

sphere on the top of fully composite pylon. The provided red zone against lightning stroke currents of 10 kA and 50 kA are displayed in Fig. 6.15, which show that all phase conductors are below the relevant spheres.

As a result, direct lightning strokes higher than 8.2 kA can be effectively protected by shield wires and the lighting stroke currents between 5 kA and 8.2 kA have lower chance to hit the upper phase conductors. Moreover, Fig. 6.12 implies the fact that only direct lightning strokes with lower magnitudes can penetrate to the center of fully composite pylon, which does not cause a concern for the insulation of pylon. Moreover, according to Figs. 6.12, 6.13, 6.14 and 6.15, it is evident that horizontal lightning strokes cannot terminate to the phase conductors on both sides of pylon, because both sides of fully composite pylon are placed in the protected zone (red zone) and the penetration of lightning strokes from these regions is unlikely. This is a strong point for fully composite pylon in comparison to other traditional pylons. Thus, the assignment of $-60°$ shielding angle for the fully composite pylon seems to be adequate and the concept of unibody cross-arm with 30° inclination angle is a feasible solution.

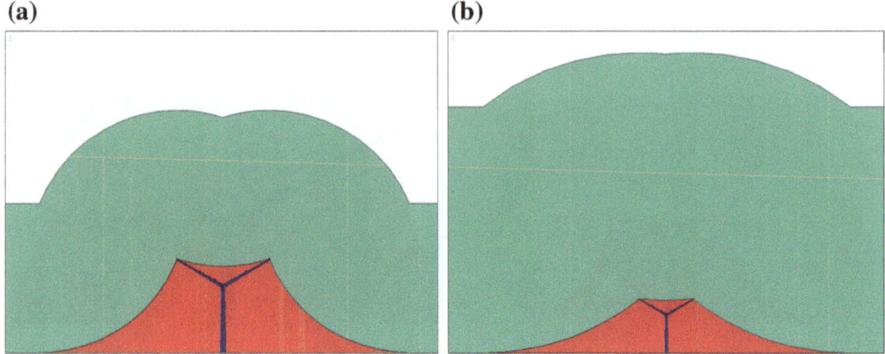

Fig. 6.15 Protected conductors in red area for lightning stroke currents of **a** 10 kA and **b** 50 kA

6.6 Summary

In this chapter, the lightning shielding performance of fully composite pylon was investigated. Since the fully composite pylon has a different configuration in comparison to other traditional pylons, the conventional electro-geometric model (EGM) cannot be directly used to evaluate the fully composite pylon's lightning protection behavior. For this reason, conventional EGM is modified to be used for the application of fully composite pylon. The assigned shielding angle of −60° for the basic design of fully composite pylon is evaluated and the results show that there is an unprotected zone at the center of the pylon against vertical lightning strikes. Therefore, instead of maximum shielding failure current in conventional EGM, minimum intersection current is used in modified EGM method to compute shielding failure rate (SFR) of the pylon. The obtained results for the SFR of both shielding angles show that shielding failure rates of fully composite pylon are within acceptable range. Moreover, shielding failure flashover rate (SFFOR) of the pylon is zero theoretically. It proves that fully composite pylon provides reliable lightning shielding performance against vertical lightning strikes.

To investigate the possible occurrence of horizontal lightning, rolling sphere method (RSM) is used to assess the protection of phase conductors against lightning strikes. The visualized results show that phase conductors on the fully composite pylon (with diagonal configuration) are fully protected against horizontal lightning strikes. The protection of phase conductors against vertical lightning is also visualized which shows that the lightning protection of fully composite pylon increases by increasing lightning stroke current above 5 kA and provides a full protection for 8.2 kA. The full protection based on the modified EGM method occurs for the lightning stroke currents higher than 3.1 kA. The discrepancy in the threshold lightning stroke currents comes from the differences in the relevant equations for striking distances. However, both approaches of the modified EGM and the RSM show a full protection for the fully composite pylon against the critical current of 8.2 kA, which in the case of occurrence, can produce an overvoltage equal to the critical flashover voltage for the insulation of unibody cross-arm and thus, can damage the cross-arm of fully composite pylon.

References

1. T. Jahangiri, *Electrical Design of a New, Innovative Overhead Line Transmission Tower Made in Composite Materials* (Aalborg University, 2018)
2. P.M. Miguel, D.M. Correia, A.C. Carvalho, Application of the leader progression model to evaluate the lightning performance of AC and DC EHV transmission lines, in *CIGRE Session Paris* (2016)
3. IEEE Std 1243-1997: IEEE Guide for Improving the Lightning Performance of Transmission Lines (1997)
4. CIGRE Working Group 01- Study Committee 33: Guide to Procedures for Estimating The Lightning Performance Of Transmission Line (1991)
5. CIGRE Working Group C4.26: Evaluation of Lightning Shielding Analysis Methods for EHV and UHV DC and AC Transmission Lines (2017)
6. D. Olason, T. Ebdrup, K. Pedersen, F.F. Da Silva, C.L. Bak, A comparison of the lightning performance of the newly designed Eagle pylon and the traditional Donau pylon, based on tower geometry, in *International Colloquium on Lightning and Power Systems (CIGRE)* (2014)
7. IEEE Std 1313.2: IEEE Guide for the Application of Insulation Coordination (1999)
8. IEC 60071-2: Insulation co-ordination—Part 2: Application guide (1996)
9. A.H. Hileman, *Insulation Coordination for Power Systems* (CRC Press, Taylor & Francis Group, 1999)
10. P.N. Mikropoulos, T.E. Tsovilis, Lightning attachment models and maximum shielding failure current of overhead transmission lines: implications in insulation coordination of substations. IET Gener. Transm. Distrib. **4**(12), 1299–1313 (2010)
11. P.N. Mikropoulos, T.E. Tsovilis, Lightning attachment models and perfect shielding angle of transmission lines, in *Proceedings of the 44th International Universities Power Engineering Conference (UPEC)* (2009)
12. T. Horvath, The protected space proved to be an undefined term, in *International Conference on Lightning Protection (ICLP)* (2012)
13. IEEE Std 998-2012: IEEE Guide for Direct Lightning Stroke Shielding of Substations (2012)
14. T. Jahangiri, C.L. Bak, F.F. da Silva, B. Endahl, J. Holbøll, Assessment of lightning shielding performance of a 400 kV double-circuit fully composite transmission line pylon, in *CIGRE Session, Paris, France*, C4 - 205 (2016)
15. D. Machidon, M. Istrate, M. Guşă, M. Dragomir, Rolling sphere method application for HV lines, in *3rd International Conference on Modern Power Systems MPS 2010* (2010)
16. N. Szedenik, Rolling sphere—method or theory? J. Electrostat. **51–52**, 345–350 (2001)
17. A. Khodadadi, M.H. Nazari, S.H. Hosseinian, Designing an optimal lightning protection scheme for substations using shielding wires. Eng. Technol. Appl. Sci. Res. **7**(3), 1595–1599 (2017)
18. P. Chowdhuri et al., Parameters of lightning strokes: a review. IEEE Trans. Power Deliv. **20**(1), 346–358 (2005)

Chapter 7
Lightning Shielding Failure Investigation by High Voltage Experiments

7.1 Introduction

Overhead lines at voltage higher than 110 kV are equipped with one or two shield wires that are grounded through the tower body or other methods at the transmission tower top, which aim to collect lightning strokes to avoid they terminating on phase conductors and leading to flashover or causing insulation breakdown issues in a substation if the stroke point is close enough to the substation. The relative position of shield wires to phase conductors are critical for shielding. This relative position is indicated by the definition of shielding angle that is specified as 'the included angle between a line connecting the shield wire with the shielded conductor and the vertical line through the shield wire' [1].

Although shield wires are applied, a 100% protection for phase conductors from direct lightning termination is impossible, unless they are surrounded by shield wires. The phenomenon that a direct lightning stroke terminates on a phase conductor, instead of on shield wires, is called 'shielding failure'. The shielding failure rate is an indicator for lightning protection performance of shield wires in transmission towers.

7.1.1 Electro-Geometric Model (EGM)

As discussed in Chap. 6, in order to evaluate the shielding failure rate for overhead lines, electro-geometric model (EGM) is proposed, which utilizes the striking distance definition as

$$r = AI^b \tag{7.1}$$

where r represents the striking distance of grounded object in meters; I represents lightning current in kA; A and b are constants.

T. Jahangiri et al., *Electrical Design of a 400 kV Composite Tower*, Lecture Notes in Electrical Engineering 557,
https://doi.org/10.1007/978-3-030-17843-7_7

With the help of electro-geometric model (EGM), the maximum lightning current that leads to shielding failure, shielding failure rate (SFR) and shielding failure flashover rate (SFFOR) for overhead lines installed in a specific transmission tower can be obtained. With these date, the lightning protection performance of shield wires in these towers can be evaluated.

7.1.2 Scale Model Test

7.1.2.1 Concept of Scale Model Test

Electro-geometric model (EGM) method is widely used for predicting the lightning shielding performance in the design of overhead lines in industry.

However, researchers have pointed out in some cases that the real shielding failure performance of overhead lines in service differs from that predicted by electro-geometric model (EGM), especially in the case of Ultra Higl Voltage (UHV) lines [2, 3]. This is because electro-geometric model (EGM) only relates the striking distance to lightning current and ignores physical process of a lightning strike. Thus, the theoretical analysis by electro-geometric model (EGM) needs to be verified by practical experience, or more efficiently, by experimental methods.

Scale model test has become an important experimental method for lightning shielding failure investigation in overhead lines from 1960s [4]. The basic concept of a scale model test is to simulate a lightning stroke terminating on reduced-scale overhead lines [5].

A lightning flash develops in steps from the thunder cloud towards the ground, instead of going straight forward [6]. This progress is figuratively named as 'stepped leader'. It is impossible to simulate the whole lightning process in a high voltage lab. Instead, researchers believe that the final stage of a lightning flash, which determines the final strike point, can be well simulated by a high voltage impulse. Thus, the lightning stroke to grounded objects can be simulated by the electrical discharge in a long air gap with high voltage impulse. This is the basis of scale model tests. To better understand the concept of scale model tests, two examples are given as follows.

S. Taniguchi found out that the measured number of direct lightning strokes terminating on phase conductors for 500 kV AC transmission lines differed from that predicted by electro-geometric model (EGM) in design stage [7]. Thus the researcher conducted scale model test to investigate lightning shielding performance of Ultra Higl Voltage (UHV) transmission lines. Equivalent conductors of typical Ultra Higl Voltage (UHV) overhead lines were installed with a scale ratio of 1 : 20, as it shows in Fig. 7.1. A metallic electrode, representing the downward leader of the lightning strike, is horizontally located beside the conductors, indicating horizontal lightning flashes. S. Taniguchi applied switching impulses between the electrode and equivalent conductors, since he supposed that the switching impulse's front time was more close to the pause time between steps of the downward leader, which was usually considered between 20 and 100 μs. Additionally, negative impulses were utilized

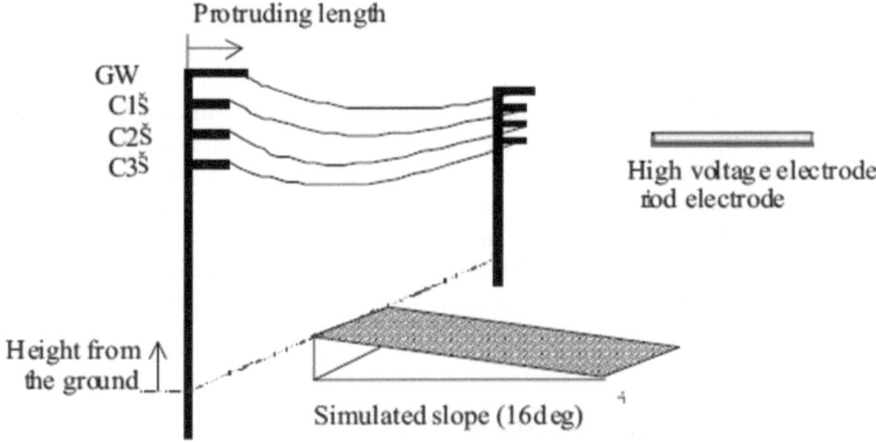

(a) Scale model test arrangement for UHV transmission lines

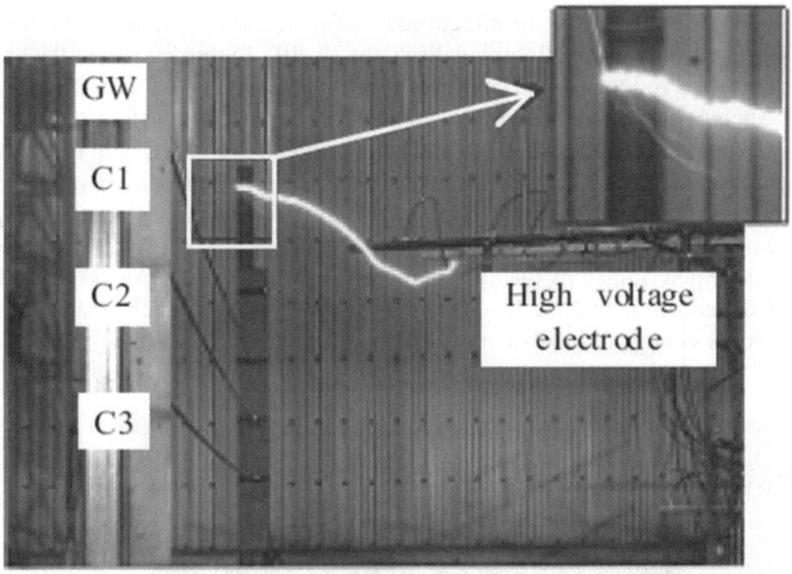

(b) An example of discharge to equivalent shield wires in the scale model test

Fig. 7.1 An example of scale model test for Ultra Higl Voltage (UHV) AC transmission lines [7]. **a** Scale model test arrangement for UHV transmission lines, **b** An example of discharge to equivalent shield wires in the scale model test

because most of lightning strikes have a negative polarity. High speed camera was utilized to capture discharge paths between the electrode to single equivalent phase conductor or ground wires. As a result, the shielding failure phenomenon with Ultra Higl Voltage (UHV) transmission lines were investigated experimentally.

For HVDC transmission lines, scale model tests are also applicable. In [8], H. He conducted scale model test to study the shielding failure characteristics of ±500 kV HVDC transmission lines. In order to guarantee the reliability of scale model test, the researcher discussed effects of the scale ratio, phase conductor's potential and electric field around the grounded object. Through scale model tests, the author found out the shielding failure rate increases with consideration of operational voltage at DC conductors, which is usually ignored in electro-geometric model (EGM).

7.1.2.2 Effectiveness of Scale Model Tests

There is a room for argument about the effectiveness of scale model test in the lightning shielding failure investigation for real overhead lines.

G. Qian believes that the effectiveness of scale model test lies in the similarity of long gap discharge and lightning discharge [9, 10]. Figure 7.2 shows the breakdown process in a rod to rod gap captured by a high speed camera with 450000 fps [10]. The height of the grounded rod is 0.5 m and the gap length is 6 m.

Compared the rod to rod gap breakdown shown in Fig. 7.2 with the last step of a lightning discharge, it can be seen the similarity in two cases. Firstly, the breakdown process in both long gap discharges and lightning discharges includes development of downward/upward leaders and following streamers. Also, the striking point in both cases are determined by the development of upward leaders from grounded objects. The difference is that the length of the upward leader is much smaller in long gap discharges compared with that in lightning discharges. It is also described in CIGRE

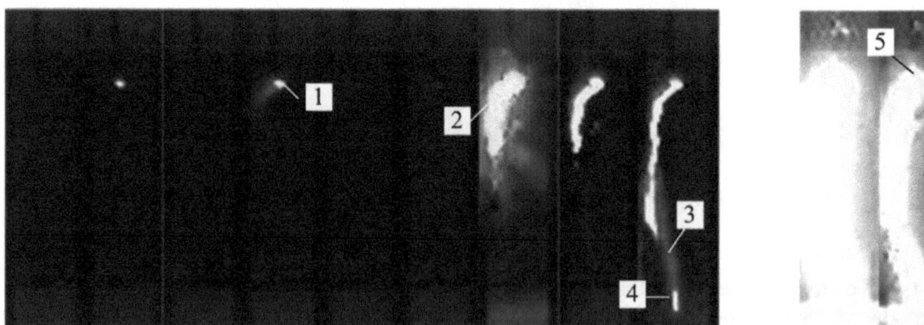

1 – Aborted negative downward leader; 2 – Negative downward leader; 3 – Bridged by streamer zone; 4 – Positive upward leader; 5 – Discharge chennel.

Fig. 7.2 Breakdown process between a 6 m rod-rod gap under standard switching impulse [10]

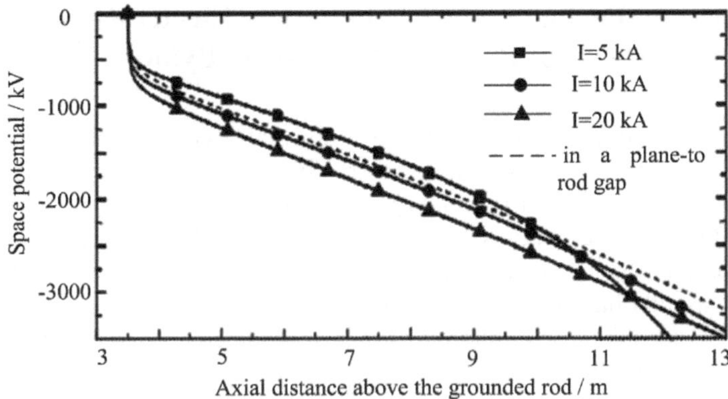

Fig. 7.3 Space potential around grounded objects with a lightning strike and in a 13 m plane-to-rod gap [10]

Technical Brochure that the final step of the downward leader of a lightning stroke can be qualitatively equivalent to the electric breakdown in a scale model test [11].

The upward leader inception electric field from the scaled model in a scale model test and real grounded objects in a lightning discharge process only depends on the spatial electric field distribution around the object, instead of on the gap length or the type of electrodes [9, 10].

The comparison has been performed between the space potential around a grounded rod in a plane-to-rod long gap with a length of 13 m and that around a grounded lightning rod when a lightning strike is approaching [10]. Figure 7.3 shows the space potential above the lightning rod with a lightning current 5, 10 and 20 kA and that above the grounded rod when a negative switching impulse with a front time of 350 μs and a magnitude of 3.2 MV is applied to the plane electrode. It can be seen from Fig. 7.3 that in both cases the space potential are similar. Thus, it is believed that the spatial electric field distribution in a scale model test is similar to that in a lightning discharge.

Effects on the credibility of a scale model test have been discussed in [4, 12]. The author pointed out that the scale ratio, grounding methods of the scaled conductors in the scaled pylon, phase conductor potential, polarity of the HV electrode and impulse waveforms play a role in the credibility of scale model test.

Although a scale model test cannot accurately simulate the lightning strikes to overhead lines, it is still regarded as a helpful experimental method in the investigation of lightning performance of overhead lines.

7.2 Shielding Performance Evaluated by Electro-Geometric Model (EGM) of the Fully Composite Pylon

The innovative fully composite pylon has a distinctive configuration, which is different from that of traditional lattice towers. Two shield wires are designed to be clamped at tips of the cross-arm and they are arranged in a straight line with the three phase conductors. According to the definition of shielding angle, the fully composite pylon has a negative shielding angle α, which is distinctive from traditional towers that usually have positive shielding angles. Due to the unusual shield wires' arrangement, the shielding failure performance in the composite tower needs to be investigated in-depth.

Electro-geometric model (EGM) method proposed by IEEE T&D Committee has been utilized in Chap. 6 to investigate the lightning performance of the fully composite pylon. According to the Electro-geometric model (EGM) method, a shielding failure zone exists in the fully composite pylon center, which is different from that in traditional lattice towers, as Fig. 7.4 shows.

Additionally, the maximum lightning current that can lead to shielding failure in the fully composite pylon is calculated as $I_m = 3.102$ kA. The shielding failure rate (SFR) and shielding failure flashover rate (SFFOR) in the fully composite pylon has been calculated as 0.0008 flashes per year for 100 km lines and 0, respectively.

Based on these results, the conclusion can be drawn that shield wires in the fully composite pylon can provide acceptable lightning protection performance for overhead lines. However, the conclusion must be verify by experimental methods.

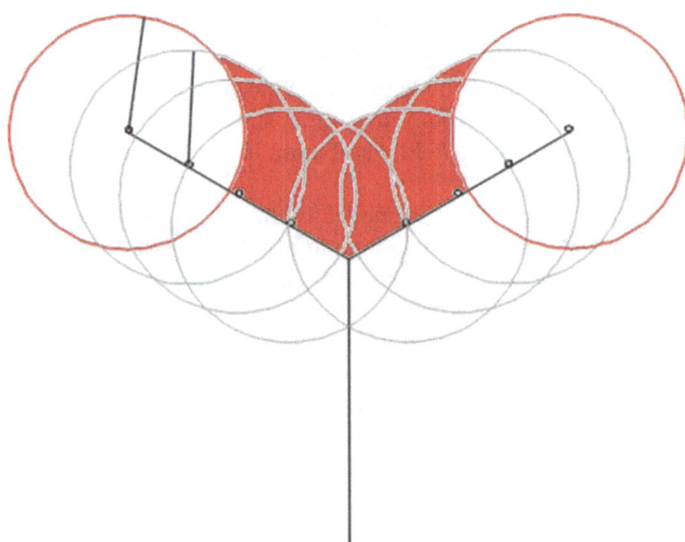

Fig. 7.4 Shielding failure zone (red zone) in the fully composite pylon based on electro-geometric model (EGM) method [5]

7.3 Scale Model Test for the Fully Composite Pylon

In order to verify the results obtained by electro-geometric model (EGM) methods and evaluate the lightning shielding performance of the fully composite pylon which has an usual negative shielding angle, high voltage experiments were conducted in the high voltage lab at Department of Energy Technology, Aalborg University, Denmark.

7.3.1 Experimental Setup

The basic idea of a scale model test is to simulate the final step of a lightning flash by an electrical discharge with an impulse voltage, which is applied to a gap between a high voltage electrode and scaled conductors installed in a reduced-scale pylon model [5].

7.3.1.1 Tower Model

In the present chapter, two scale models of the fully composite pylon were applied. They were made from polyurethane, thus the models were non-conductive as the real composite pylon, which was intended to be made from Fiberglass Reinforced Plastic (FRP) materials. A scale ratio of 1 : 40 was adopted, thus the total height of the pylon model was 56 cm approximately. The inclined angle of the cross-arm from the ground plane has effects on shielding performance, since it determines the shielding angle of the pylon. In order to verify these effects, two inclined angles, 20° and 30° were adopted. As a result, the pylon model was in a 'X' shape to contain two different inclined angles. Figure 7.5 displays the pylon model configuration. It can be seen that there were several holes in the model cross-arm which were used for fixation of scaled conductors.

The air clearances in the model were referred to the minimum values determined in [13] with a scale ratio of 1 : 40, i.e.

$$\begin{cases} D_{pp}/d_{pp} = 3.6\ \text{m}/9\ \text{cm}; \\ D_{el}/d_{el} = 2.8\ \text{m}/7\ \text{cm} \end{cases} \tag{7.2}$$

where D_{pp}/d_{pp} represents distance between adjacent phase conductors in the real pylon and in the pylon model respectively. While D_{el} and d_{el} represents distance between the phase conductor and the shield wire in the real pylon and in the pylon model respectively.

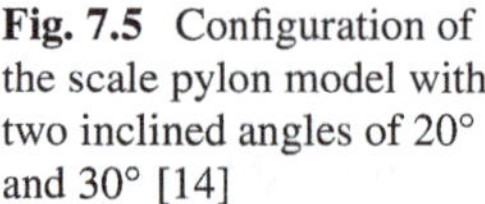

Fig. 7.5 Configuration of the scale pylon model with two inclined angles of 20° and 30° [14]

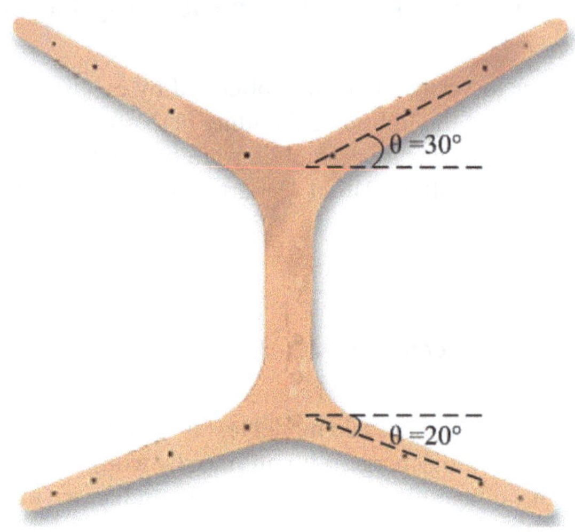

7.3.1.2 Scaled Conductors

Single steel solid wires were adopted to represent scaled phase conductors in the pylon model. The wires had a radius of 3 mm, thus in fact it represented the real bundle phase conductors with an equivalent radius of 120 mm. In reality, diameters of shield wires are usually smaller than that of phase conductors to increase the possibility of lightning attachment to shield wires. However, in present chapter, the scaled shield wires possess the same radius of 3 mm with that of the scaled phase conductors in order to have the same sagging degree. In this case, effects of sagging can be eliminated in the scale model tests. All conductors were inserted into the holes in the cross-arm, which was different from the real case that all conductors were clamped onto cross-arms by well-designed cable clamps. However, since the lightning flash was designed to happen in the middle span in the test, the different fixation method for scaled conductors in the model did not have great effects on the final results.

The grounding methods of scaled conductors in scale model test is controversial. In [12], different methods by researchers are introduced and summarized. Some researchers do not ground the scale conductors and leave them open in the test. However, the other researchers posses the opinion that scaled conductors should be grounded through a resistor whose resistance value is equivalent to that of the line surge impedance. They believe that the grounding resistance can restrict the progression velocity and amount of charges in the channel of the upward leader from the grounded object, thus a lower lightning termination probability will be obtained which is more realistic. They also found out that with the increase of grounding resistance, the lightning termination probability decreases.

In the present chapter, the scale phase conductors are grounded through a resistor with the resistance value of 300 Ω. The grounding resistance is larger than the line surge impedance $Z_c = 138.4\ \Omega$ [15] in the fully composite pylon due to equipment limitation in the lab. However, the dimension of the fully composite pylon and the pylon model is relative small, thus gap lengths in the scale model test is rather small (approximately within 1 m). With these rather small gap lengths, upward leaders' characteristics from scale phase conductors have less impacts on the final lightning termination paths. Thus the selection of 300 Ω grounding resistance is reasonable in the present case. Differing from scale phase conductors, the two scale shield wires were grounded directly.

It is also controversial that whether the potential at phase conductors should be considered or not in the scale model test. Some researchers hold the opinion that phase conductors' potential has little influence on the electric field distribution around overhead lines when a lightning flash approaches [12, 16]. While some researchers believe the phase conductors' potential cannot be neglected in a scale model test, especially for Ultra High Voltage (UHV) transmission lines [8]. In the present chapter, potential at phase conductors were not considered for simplicity.

7.3.1.3 HV Electrode

It is the final stage of the downward leader that determines the striking path, thus only the final stage of the downward leader was simulated which is reasonable. Thus, a steel solid rod, i.e. the HV electrode, with a diameter of 10 mm and length of 2 m was applied to simulate the final stage of the downward leader. The rod has an hemispheric cap. As the downward stepped leader is usually considered developing towards the earth vertically, the steel rod was adopted and placed vertically above the scale model. The electrode was connect to the output of an eight stages Marx impulse generator with characters of 800 kV and 24 kJ.

The experimental setup and the test arrangement in the lab is shown in Fig. 7.6. The HV electrode is in boldface for better visibility. The length of the midspan is 3 m.

7.3.1.4 Impulse Waveforms

The impulse waveforms' characteristics, including front time and polarity have great effects on the breakdown voltage of a specific gap thus on discharge paths.

According to categorization of lightning types in [17], negative downward flash is dominant which covers 85–90% of lightning flashes, especially for structures with moderate height. Thus, in the present chapter, a negative impulse was applied.

Determination of front time of the impulse in the scale model test is controversial. Some researchers believe the standard lightning impulse defined by IEC 60071 with a front/tail time of $1.2/50\ \mu s$ should be applied [18]. While other researchers believe the front time should be in accordance with the pause time between steps of the downward stepped leader [8, 19]. Since the pause time ranges from 10 to 100 μs, they believe it is more realistic to apply a switching impulse (slow-front impulse)

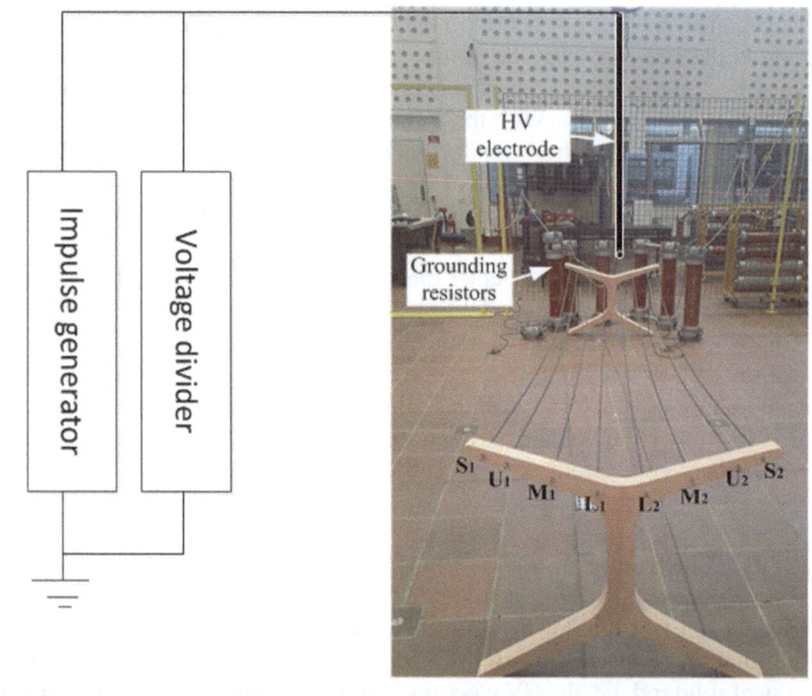

Fig. 7.6 Scale model test setup and arrangement: S_X are scale shield wires, while U_X, M_X and L_X are scale upper phase, middle phase and lower phase conductors respectively [5]

instead of a lightning impulse (fast-front impulse) in a scale model test. What's more, some researchers found out the results with a fast-front impulse and with a slow-front impulse do not differ a lot from each other, thus either one can be adopted [20, 21].

In the present chapter, both fast-front time impulse and slow-front time impulse were applied in order to obtain a reliable conclusion. Figure 7.7 shows the impulses applied in the scale model test, which were measured by an oscilloscope.

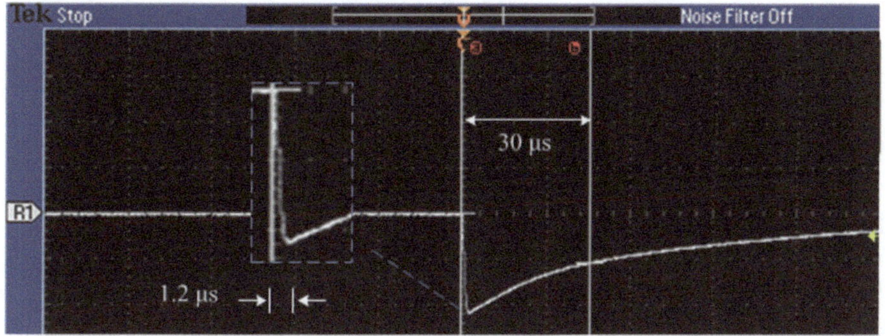

(a) Impulse voltage waveform: -1.2±30%/50±20% μs

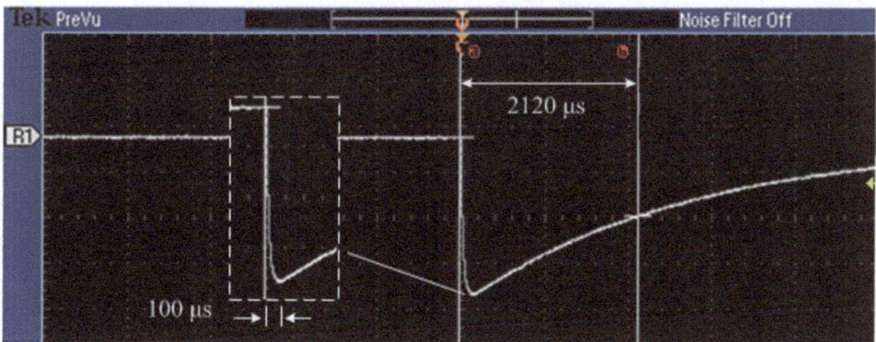

(b) Impulse voltage waveform: -100±30%/2500±20% μs

Fig. 7.7 A fast-front impulse (**a**) and a slow-front impulse (**b**) applied in the scale model test [14]

7.3.2 *Test Progress*

In the scale model test, the position of the high voltage electrode varies to get different striking distances, which represents lightning flashes carrying different lightning currents.

To better express the electrode's position, a coordinate system was established, as Fig. 7.8 indicates. The xy plane was perpendicular to the ground plane. The origin was set at the midpoint of the connecting line between S_1 and S_2. The directions of x-axis and y-axis are shown in Fig. 7.8. For simplicity, along the midspan direction, the electrode was fixed at the midpoint of the connecting line of two models. Thus, the position of the electrode could be expressed uniquely by its coordinates $(x,\ y)$. In reality, the electrode's position only changed in the red region in Fig. 7.8a, i.e. in the first quartile of the coordinate system, since the setup was axial symmetric. Table 7.1 shows positions of the electrode applied in the test.

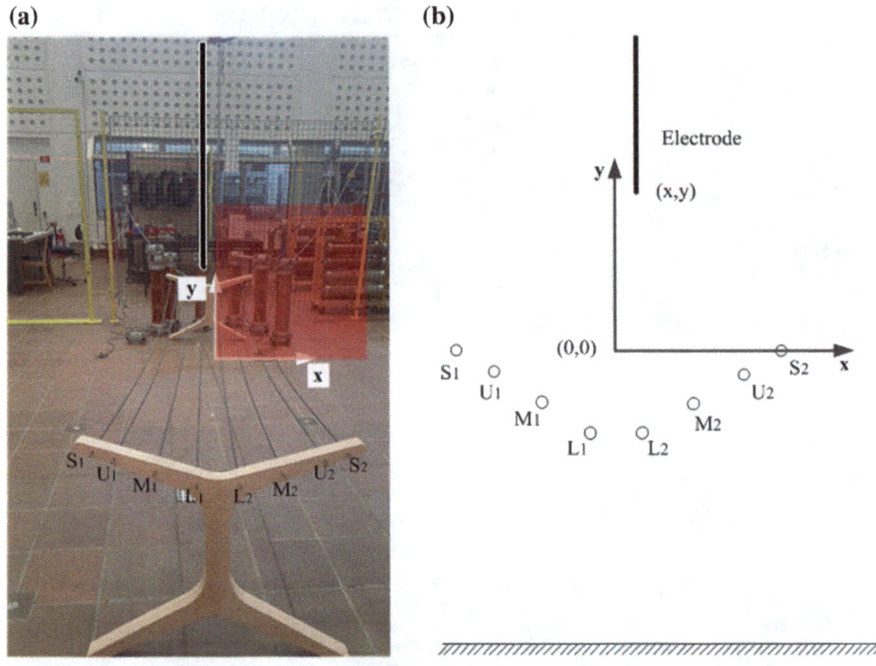

Fig. 7.8 Coordinate system in the space (left) and the 2D coordinate system (right) [5]

Table 7.1 HV electrode positions

(x, y)		x [cm]				
		0	3	5	7	10
y [cm]	15	(0, 15)	(3, 15)	(5, 15)	(7, 15)	(10, 15)
	20	(0, 20)	(3, 20)	(5, 20)	(7, 20)	(10, 20)
	25	(0, 25)	(3, 25)	(5, 25)	(7, 25)	(10, 25)
	30	(0, 30)	(3, 30)	(5, 30)	(7, 30)	(10, 30)
	35	(0, 35)	(3, 35)	(5, 35)	(7, 35)	(10, 35)
	40	(0, 40)	(3, 40)	(5, 40)	(7, 40)	(10, 40)
	45	(0, 45)	(3, 45)	(5, 45)	(7, 45)	(10, 45)

For each (x, y), a pre-test was done. The breakdown voltage $U_{90\%-100\%}$, which was defined as a voltage that can lead to breakdown of a specific gap with a probability of 90–100%, was obtained by test in advance for each electrode position. The pre-test contained the following steps:

1. Raise the voltage U from zero at a constant rate of 10 kV/s until breakdown happens to the gap; the breakdown voltage is recorded as U_0;

2. Apply impulse U with a peak value of U_0 to the gap for 10 times; if more than 8 times of breakdown happens, then U_0 is recorded as $U_{90\%-100\%}$; otherwise, adopt the third step;
3. Apply impulse U with a peak value of $U_1 = 1.1U_0$ to the gap for 10 times; if more than 8 times of breakdown happens, then U_1 is recorded as $U_{90\%-100\%}$; if not, increase the peak value of the impulse by 10%, until more than 8 times of breakdown happens to the gap with the same impulse and record the peak value as $U_{90\%-100\%}$ for the specific gap.

For each high voltage electrode position, an impulse voltage with a peak value of $U_{90\%-100\%}$ was applied for N times and $N = 50$. The discharge paths were recorded by a digital camera, which was placed outside of the metallic fence that surrounding the test setup. The camera could take 1250 pictures per second. Thus the number N_{SF} of discharges happening to scale phase conductors were recorded. The shielding failure rate (SFR) was calculated by

$$SFR = \frac{N_{SF}}{N} \times 100\% \tag{7.3}$$

7.3.3 Test Results and Analysis

Since the initial designed inclined angle of the cross-arm in the fully composite pylon is 30°, all results presented in Sect. 7.3.3 are obtained with the 30° inclined angle, i.e. with the −60° shielding angle. The effects of different inclined angles will discussed later.

- Discharge paths

Figures 7.9 and 7.10 show several discharge paths taken by the digital camera with a fast-front impulse and a slow-front impulse, respectively. Figures 7.9b, c and 7.10a display that the discharge happened to scale phase conductors, i.e. shielding failure happened. While in the rest photos, discharges happened to scale shield wires, where scale phase conductors were protected.

It is within expectation that discharge happened at a gap where the shortest gap length exists, such as in Fig. 7.9a, b. However, there were exceptions that discharge happened to longer gaps instead of always to the shortest one. For example, in Fig. 7.9c, the discharge happened to L_1 with the fact that the shortest gap existed between the high voltage electrode and M_2. Also, in Fig. 7.9d, the discharge should have happened to scale phase conductors. As a matter of fact, it happened to the scale shield wire S_1. It is similar in Fig. 7.10b, c, d, where discharges happened to longer gaps.

To explain these exceptions, two facts should be considered: (1) the difference in gap lengths in the pylon model is not significant; (2) scale phase conductors were grounded through equivalent surge impedances while scale shield wires were grounded directly.

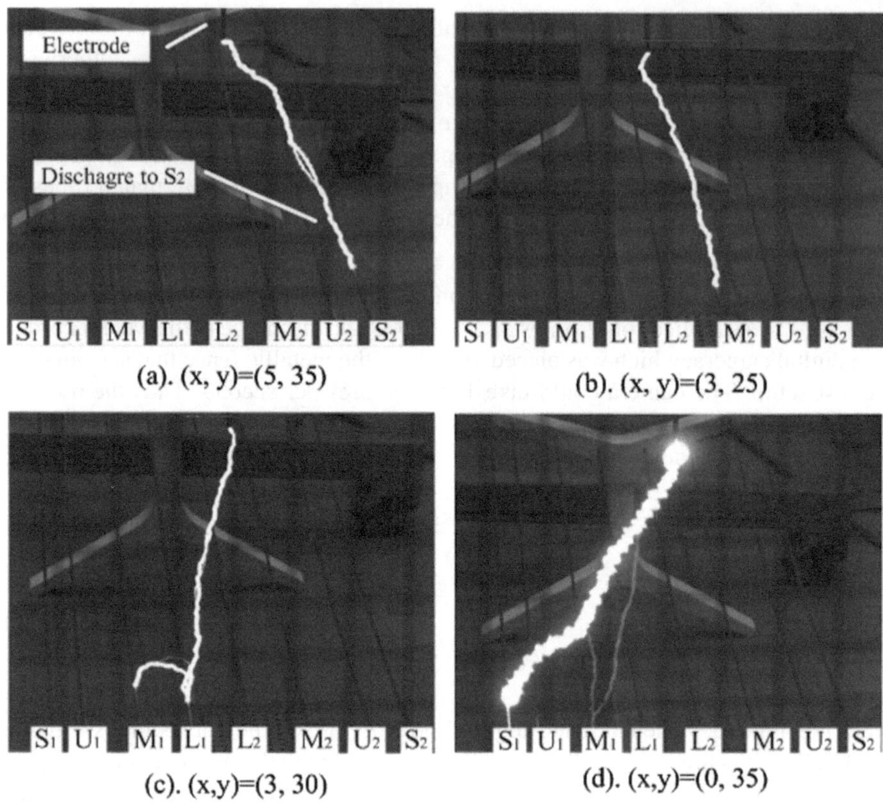

Fig. 7.9 Discharge paths with impulses waveform of $-1.2 \pm 30/50 \pm 20\ \mu\text{s}$ [5]

For the first factor, due to a small difference which was smaller than 5 cm in gap lengths in the scale model test, the natural randomness of discharge with impulse waveforms have greater effects. Thus discharges to longer gaps were observed. While for the second factor, the equivalent surge impedance impeded charge accumulation on scale phase conductors, thus the development of upward leader was delayed from scale phase conductors compared with that of scale shield wires, which were grounded directly. As a result, the discharge was attracted by scale shield wires, such as in Figs. 7.9d and 7.10b, c, d.

Flashover between phases were observed in Figs. 7.9c and 7.10a. Meanwhile, weak discharge branches were observed from the main discharge channel such as in Figs. 7.9d and 7.10d.

It should be noticed that, among all discharges in the test, no single discharge ever happened to U_1 and U_2. It is attributed to the small air clearance between the scale upper phase conductor and the scale shield wire, which is only 7 cm approximately [5].

More photos taken in the scale model test are referred to Appendix B.

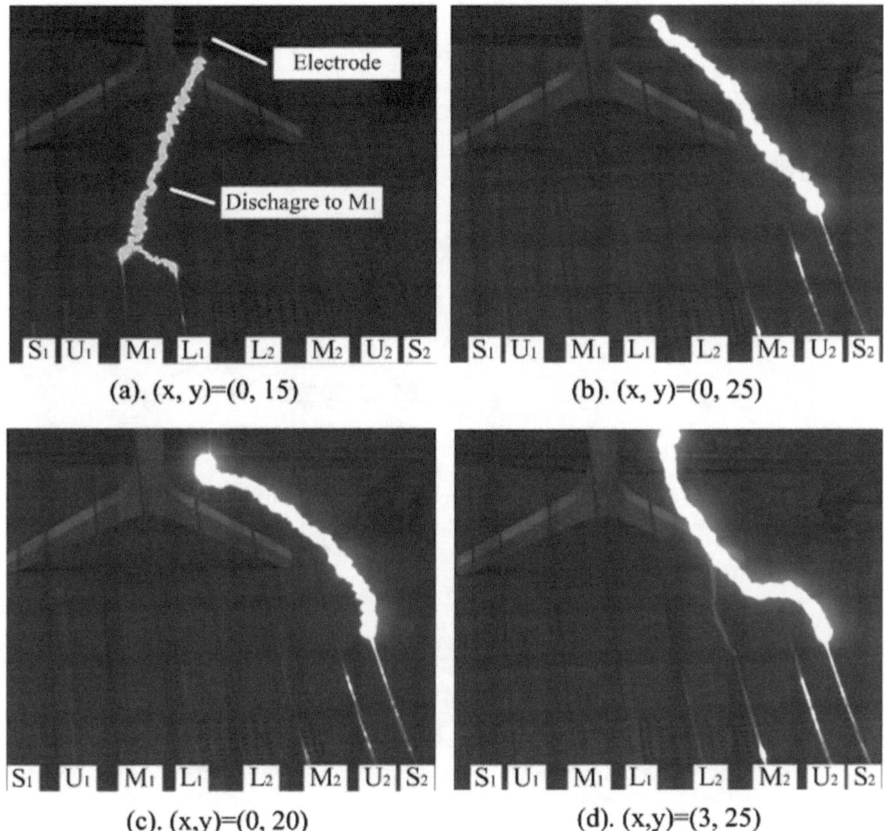

Fig. 7.10 Discharge paths with impulses waveform of $-100 \pm 30\%/2500 \pm 20\%$ μs [5]

- Shielding failure rate (SFR)

Figure 7.11 displays the shielding failure rate (SFR), calculated with Eq. (7.3), for different electrode positions with two impulse waveforms.

It is observed that when the x position was fixed, the value of shielding failure rate (SFR) decreased with the increase of y, i.e. with the increase of the electrode height. There were several small fluctuations, for example, the value of shielding failure rate (SFR) at (0, 35), which was 24%, was bigger than that at (0, 30) which was 22% in Fig. 7.11a. The similar phenomenon was also observed in Fig. 7.11b when $x = 0$ and $x = 5$. However, the principle trending was fulfilled in both Fig. 7.11a and b that the increase of electrode height reduced the possibility of shielding failure, if neglecting those small fluctuations

While when the y position was fixed, the value of shielding failure rate (SFR) decreased with the increase of x, i.e. with the increase of horizontal distance between the electrode and the pylon center, until shielding failure rate (SFR) was reduced to

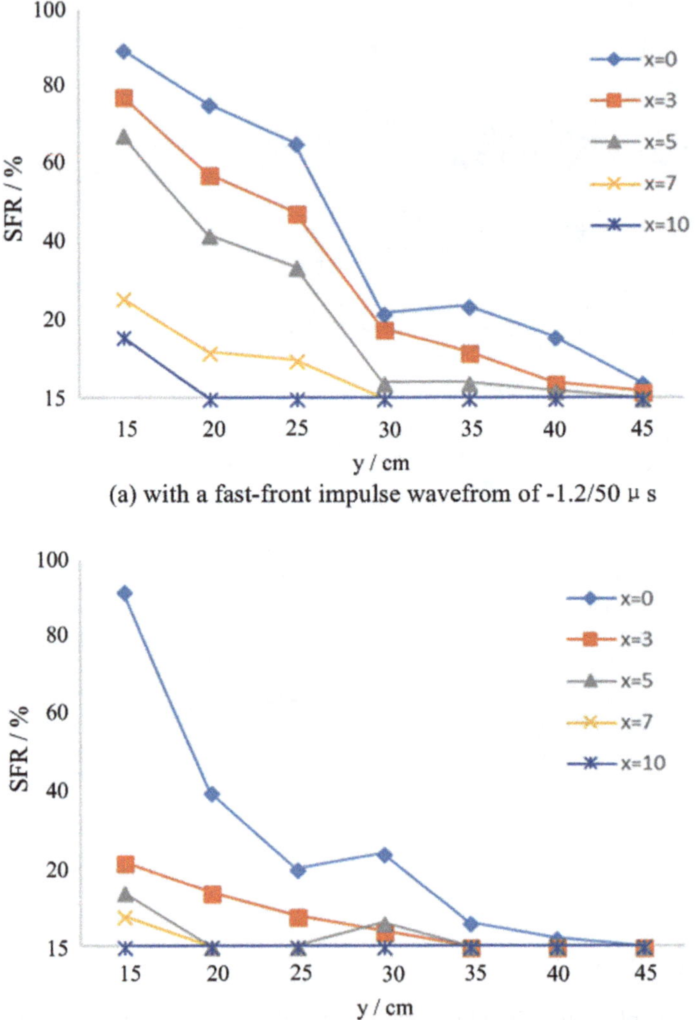

(a) with a fast-front impulse wavefrom of -1.2/50 μ s

(b) with a slow-front impulse wavefrom of -100/2500 μ s

Fig. 7.11 Shielding failure rate (SFR) for different electrode positions with different impulse waveforms

zero. There was only one exception in Fig. 7.11b that when $y = 30$, the value of shielding failure rate (SFR) at $x = 5$, which was 6%, was bigger than that at $x = 3$, which was 4%.

The two trending mentioned above can be explained that when the value of x increased from zero or when the value of y increased from 15, the gap length between the electrode and shield wire decreased. Thus, the possibility of shielding failure increased.

It can also be observed that the front-time had effects on the value of shielding failure rate (SFR) from Fig. 7.11. With the same electrode position, i.e. the same $(x,\ y)$, the value of shielding failure rate (SFR) was bigger with fast-front impulse compared with that with slow-front impulse. For example, the value of shielding failure rate (SFR) was 78% at (0, 15) with the fast-front impulse, while it was 22% at the same position with the slow-front impulse.

The possible explanation for the phenomenon might be that, with longer front-time, the downward streamer developed from the electrode could travel further towards the scale conductors before the final discharge happened. In another word, the effective striking distance was smaller with a slow-front impulse compared with that with a fast-front impulse. With a shorter striking distance, the effects of charge accumulation velocity on the final discharge paths were more significant, i.e. the downward streamer would be attracted earlier by conductors on which charge accumulation happened faster thus upward streamer developed fast. As mentioned before, the charge accumulation was faster on scale shield wires due to grounding directly, the discharges were more likely to terminate on shield wires.

7.4 Comparison of Electro-Geometric Model (EGM) and Scale Model Test Results

7.4.1 Shielding Failure Zone

Based on electro-geometric model (EGM) method, the shielding failure zone in the fully composite pylon is found in the middle of the pylon, as the red region in Fig. 7.4 shows.

Figure 7.11 also indicates that the shielding failure probability decreases to zero when the electrode position increases vertically or horizontally. Figure 7.12 displays the spatial shielding failure zone around the pylon model. P_{sf} inidates shielding failure rate in Fig. 7.12. Since the electrode positions only varied in the first quadrant and the pylon was axial symmetrical, the values of the shielding failure rate (SFR) in the second quadrant were set as identical to that at the symmetrical positions in the first quadrant.

It's obvious from Fig. 7.12 that the shielding failure zone is in the middle of the pylon. If the electrode position moves further vertically and horizontally, the shielding failure rate (SFR) will become zero. The conclusion obtained by scale model test is consistent with that obtained by electro-geometric model (EGM) method.

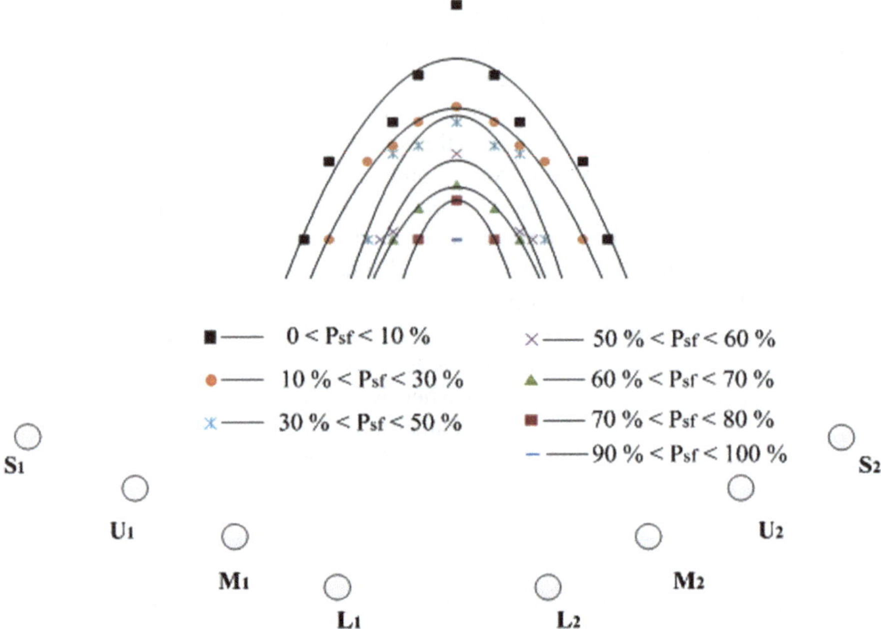

Fig. 7.12 Spatial shielding failure probability distribution zone with a fast-front impulse of −1.2/50 μs [14]

7.4.2 Maximum Shielding Failure Current

I_m is defined as the maximum lightning current when considering shielding failure. In the scale model test, the value of I_m is unknown, thus it needs to be interpreted from other parameters.

Striking distances r_c for each electrode position in the scale model test is the gap distance between the electrode and the struck conductor. For simplicity, we apply the assumption that all the gaps are in the 2D xy plane. Since the scale ratio of $1:40$ was applied in the scale model test, the real striking distances R_C can be calculated by [5]

$$R_C = 40 \times r_c \tag{7.4}$$

The corresponding values of lightning current can be obtained by substituting values of R_C into the striking distance definition equation Eq. (7.5) recommended by IEEE T&D Committee [22].

$$R_c = 10I^{0.65} \tag{7.5}$$

where I is the lightning current in kA.

Tables 7.2 and 7.3 show the lightning currents and shielding failure rate (SFR) when the electrode was located at $(0,\ y)$. The other positions are not considered since at the same electrode height, r_c is the biggest at $x = 0$ thus the corresponding current is the biggest.

It should be noted that there is a small deviation in values of r_c thus R_C in Tables 7.2 and 7.3 with the same $(x,\ y)$. For example, $r_c = 0.46$ m at (0, 25) with the fast-front impulse while the value is 0.49 m with the slow-front impulse. This is because in the scale model test, all the striking distances were adjusted and measured manually, thus an unavoidable error existed, which was smaller than 7%.

It can be figured out that when lightning current magnitude increases, the value of shielding failure rate (SFR) decreases. This conclusion is also consistent with the results obtained by electro geometric model (EGM) methods.

In the design of location and number of shield wires, a suitable value of shielding failure rate (SFR) is usually set as the acceptable criterion, considering a shielding failure rate (SFR) of zero is virtually impossible. In the present chapter, a shielding failure rate (SFR) of 5% is set as the criterion. According to Tables 7.2 and 7.3, when the electrode is located at (0, 45) and (0, 35) respectively, the shielding failure

Table 7.2 Lightning current versus shielding failure rate (SFR) with an impulse of ($-1.2/50\ \mu s$) [5]

$(0,\ y)$	r_c/R_C [m]	Lightning current [kA]	SFR [%]
(0, 15)	0.33/13.2	1.53	90
(0, 20)	0.42/16.8	2.22	76
(0, 25)	0.46/18.4	2.56	66
(0, 30)	0.50/20.0	2.90	22
(0, 35)	0.56/22.4	3.46	24
(0, 40)	0.62/24.8	4.04	16
(0, 45)	0.66/26.4	4.45	4

Table 7.3 Lightning current versus shielding failure rate (SFR) with an impulse of ($-100/2500\ \mu s$) [5]

$(0,\ y)$	r_c/R_C [m]	Lightning current [kA]	SFR [%]
(0, 15)	0.35/14.0	1.68	92
(0, 20)	0.44/17.6	2.39	40
(0, 25)	0.49/19.6	2.82	20
(0, 30)	0.53/21.2	3.18	24
(0, 35)	0.58/23.2	3.60	4
(0, 40)	0.64/25.6	4.45	2
(0, 45)	0.69/27.6	4.77	0

rate (SFR) is acceptable. Thus, values of I_m for the fast-front impulse and for the slow-front impulse are

$$\begin{cases} I_{m1} = 4.45 \text{ kA} \\ I_{m2} = 3.60 \text{ kA} \end{cases} \tag{7.6}$$

respectively according to Tables 7.2 and 7.3.

The deviation of I_{m1} and I_{m2} is not remarkable. This conclusion accords with that in [20, 21], i.e. the application of a fast-front impulse or of a slow-front impulse to a scale model test does not give a big difference in the final results.

According to the comparison of I_{m1}, I_{m2} and $I_m = 3.102$ kA which is obtained by electro-geometric model (EGM) method, the test results obtained by these two methods, i.e. experimental method and electro-geometric model (EGM) method, do not show big difference, which means these two methods support each other's application to the lightning protection design for the innovative fully composite pylon [5].

7.4.3 Shielding Failure Rate (SFR) and Shielding Failure Flashover Rate (SFFOR)

It has been proved that the statistical distribution of first stroke current magnitude can be assumed by lognormal distribution. Thus, its probability density function can be expressed by [17]

$$f(I) = \frac{1}{\sqrt{2\pi}\beta I} e^{-\frac{1}{2}\left[\frac{ln(I/M))}{\beta}\right]^2} \tag{7.7}$$

where β is the log standard deviation and M is the median.

CIGRE has recommended following values for M and β with I_F as the peak value of the first stroke [23].

$$\begin{cases} M = 61.1, \beta = 1.33 \; if \;\; I_F < 20 \text{ kA} \\ M = 33.3, \beta = 0.605 \; if \;\; I_F > 20 \text{ kA} \end{cases} \tag{7.8}$$

According to Eqs. (7.7) and (7.8), the accumulative probability that lightning current magnitude I_0 is smaller than I_{m1} and I_{m2} is

$$\begin{cases} P(I_0 \leq I_{m1}) = 0.64\% \\ P(I_0 \leq I_{m2}) = 0.37\% \end{cases} \tag{7.9}$$

respectively.

The small probability of lightning current which is smaller than I_{m1} and I_{m2} indicates that the shielding failure rate (SFR) in the fully composite pylon is very small.

What's more, the minimum lightning current when considering flashover caused by shielding failure is recommended as 3 kA [23]. It can be inferred that the shielding failure flashover rate (SFFOR) for the fully composite pylon is very small. And these conclusions are also consistent with that obtained by the electro geometric model (EGM).

According to electro geometric model (EGM) and scale model tests, it can be concluded that shield wires in the fully composite pylon can provide acceptable protection for overhead lines.

7.4.4 Effects of the Cross-Arm Inclined Angle

As is introduced in Chap. 6, the shielding failure zone in the fully composite pylon is affected by θ, the cross-arm inclined angle from the ground plan. In order to investigate the effects, two inclined angles, 30° which is the initial design inclined angle, and 20° were applied. That is, two shielding angles, −60° and −70° were applied.

According to the electro-geometric model (EGM) methods introduced in Chap. 6, the maximum shielding failure current I_m are calculated for $\theta_1 = 30°$ and $\theta_2 = 20°$, respectively.

$$\begin{cases} I_{m-\theta_1} = 3.102 \text{ kA} \\ I_{m-\theta_2} = 6.127 \text{ kA} \end{cases} \tag{7.10}$$

Table 7.4 Values of shielding failure rate (SFR) for different cross-arm inclined angles [24]

(0, y)	SFP with $\theta = 20°$ [%]	SFP with $\theta = 30°$ [%]
(0, 15)	96	92
(0, 20)	90	40
(0, 25)	78	20
(0, 30)	60	24
(0, 35)	44	4
(0, 40)	28	2
(0, 45)	32	0
(0, 50)	18	0
(0, 55)	8	0
(0, 60)	4	0

Table 7.5 Maximum shielding failure current I_m for different cross-arm inclined angles [24]

θ	Electrode location	r_c/R_C [m]	SFR [%]	I_m [kA]
20°	(0, 60)	0.85/34	4	6.57
30°	(0, 35)	0.58/23.2	4	3.60

To verify the results in Eq. (7.10), a scale model test was conducted for $\theta_1 = 30°$ and $\theta_2 = 20°$ with the pylon model shown in Fig. 7.5. The negative slow-front impulse with the waveform of $-100/2500$ μs was applied.

By using the same method introduced in Sect. 7.4.2, values of shielding failure rate (SFR) when the electrode located at $(0, \ y)$ were calculated for both inclined angles and the results are shown in Table 7.4.

Again, set $SFR = 5\%$ as the acceptable criterion, the maximum shielding failure current $I_m$with different θ is obtained in Table 7.5.

With the same cross-arm inclined angle, the maximum shielding failure current obtained by electro-geometric model (EGM) analysis and experimental method do not differ a lot, which means the two methods verify each other.

And both methods indicate that the bigger cross-arm inclined angle (30°) leads to a smaller value of I_m, which means a bigger inclined angle (30°) provides better protection from direct lightning strikes to phase conductors in the fully composite pylon [5].

7.5 Summary

This chapter has verified with experimental methods the lightning shielding performance for overhead lines in the fully composite pylon, which has a negative shielding angle due to its distinctive configuration.

In order to verify the conclusion obtained by electro-geometric model (EGM) analysis in Chap. 6, a scale model test has been performed. Direct lightning strikes to phase conductors were simulated by electrical discharges to scale phase conductors with fast-front and slow-front impulses in the lab [5]. Shielding failure rate (SFR) was obtained by interpreting the ratio of number of discharges to scale phase conductors to the total number of discharges [5]. Data in the scale model test indicates that the shielding failure zone exists in the pylon center [5]. And with the striking point moving away from the pylon center vertically or horizontally, the shielding failure rate (SFR) decreases [5]. By setting $SFR = 5\%$ as the acceptance criterion in the test, the maximum shielding failure current $I_{m1} = 4.45$ kA with the fast-front impulse and $I_{m2} = 3.60$ kA for the slow-front impulse are obtained.

Comparing results developed by theoretical method—electro-geometric model (EGM) and experimental method-scale model test, it is obvious that the deviation in both results is quite small. In fact, the two methods verify each other and support each other's application to the lightning shielding investigation in the innovative fully composite pylon [5]. Finally, a conclusion can be drawn that, as regards to protection from direct lightning strikes, the design -location and number of shield wires in the fully composite pylon fulfills requirements.

References

1. I. S. 1410, *Guide for improving the lightning performance of electric power overhead distribution lines*, Std., Jan 2011
2. S. Taniguchi, S. Okabe, T. Takahashi, T. Shindo, Discharge characteristics of 5 m long air gap under foggy conditions with lightning shielding of transmission line. IEEE Trans. Dielectr. Electr. Insul. **15**(4), 1031–1037 (2008)
3. J. Takami, S. Okabe, Characteristics of direct lightning strokes to phase conductors of UHV transmission lines. IEEE Trans. Power Deliv. **22**(1), 537–546 (2007)
4. Y. An, Y. Hu, X. Wen, Z. Jiang, L. Lan, Review of experimental study on lightning shielding performance of transmission line. High Volt. Appar. **52**(7), 1–9 (2016). (in Chinese)
5. Q. Wang, T. Jahangiri, C.L. Bak, F.F. da Silva, H. Skouboe, Investigation on shielding failure of a novel 400 kV double-circuit composite tower. IEEE Trans. Power Deliv. **33**(2), 752–760 (2018)
6. A.R. Hileman, *Insulation Coordination for Power Systems*. (Taylor & Francis, 1999). Chapter 6: The lightning flashes
7. S. Taniguchi, S. Okabe, A contribution to the investigation of the shielding effect of transmission line conductors to lightning strikes. IEEE Trans. Dielectr. Electr. Insul. **15**(3), 710–720 (2008). Jun
8. H. He, J. He, D. Zhang, L. Ding, Z. Jiang, C. Wang, H. Ye, Experimental study on lightning shielding performance of +/− 500 kV transmission lines, in *Conference on Asia-Pacific Power and Energy Engineering*, Wuhan, China, Mar 2009, pp. 1–7
9. G. Qian, X. Wang, Y. Wang, The lightning shielding simulation theory and test technology. High Volt. Eng. **24**(2), 26–31 (1998). (in Chinese)
10. W. Chen, H. He, G. Qian, J. Chen, Review of the lightning shielding against direct lightning strokes based on laboratory long air gap discharges. Proc. Chin. Soc. Electr. Eng. **32**(10), 1–12 (2012)
11. C. W. C4-26, Evaluation of lightning shielding analysis methods for EHV and UHV DC and AC transmission lines, CIGRE Technical Brochure, Oct 2017
12. Z. An, L. Lan, X. Wen, Y. Wang, Impacting factors of large sized model test for lightning shielding performance of UHV transmission lines. Power Syst. Technol. **38**(5) (2014)
13. T. Jahangiri, C.L. Bak, F.F. Silva, B. Endahl, Determination of minimum air clearances for a 420 kV novel unibody composite cross-arm, in *50th International Universities Power Engineering Conference (UPEC)*, Stokeontrent, UK, Sep 2015, pp. 1–6
14. Q. Wang, C.L. Bak, F.F. Silva, H. Skouboe, Scale model test on a novel 400 kV double-circuit composite pylon, in *International Conference on Power System Transients* Seoul, South Korea, Jun 2017
15. T. Jahangiri, C.L. Bak, F.F. Silva, B. Endahl, Assessment of lightning shielding performance of a 400 kV double-circuit fully composite transmission line pylon, in *Cigré Session*, Paris, France: CIGRE, Aug 2016
16. M.A. Uman, *The Lightning Discharge*. (Courier Corporation, 2001)
17. A.R. Hileman, *Insulation Coordination for Power Systems*. (Taylor & Francis, 1999). Chapter 6: The lightning flashes, p. 205
18. M.S. Banjanin, M.S. Savić, Z.M. Stojković, Lightning protection of overhead transmission lines using external ground wires. Electr. Power Syst. Res. **127**, 206–212 (2015)

19. T. Disyadej, S. Mallick, S. Grzybowski, Laboratory study for estimating the number of lightning flashes to transmission lines, in *North American Power Symposium (NAPS)*, Starkville, USA, Oct 2009, pp. 1–4
20. K. Miyake, I. Kishizima, T. Suzuki, Study on experimental simulation of lightning strokes. IEEE Power Eng. Rev. **4**, 41–42 (1981)
21. Y. Wang, Y. An, E. Shenglong, X. Wen, Factors effect on discharge path to scaled UHV transmission lines, in *International Conference on Lightning Protection*, Oct 2014 (IEEE, Shanghai, China, 2014), pp. 493–497
22. I.W. Group, et al., A simplified method for estimating lightning performance of transmission lines, *IEEE Trans. Power App. Syst.* **104**(4), 919–932 (1985)
23. C. W. 33-01, *Guide to procedures for estimating the lightning performance of transmission lines*. CIGRE (1991)
24. Q. Wang, T. Jahangiri, C.L. Bak, F.F. da Silva, Experimental evaluation of shielding angles effects on lightning performance in a 400 kV AC double-circuit composite pylon, in *CIGRÉ Symposium*, Dublin, Ireland, May 2017

Chapter 8
Environmental Effects of Fully Composite Pylon

8.1 Introduction

The electromagnetic environmental problems of overhead transmission lines are important issues in the planning and design of new overhead lines. Nowadays, public opinion and their opposition regarding the environmental effects of overhead lines should be taken into account by Transmission System Operators (TSOs) when refurbishing or building new overhead lines are required. These environmental problems include power frequency issues related to low electromagnetic field emission and the corona performance of the lines. Since the fully composite pylon is in the design process and there is no measured data for the corona performance of the pylon, therefore the environmental effects of the fully composite pylon is one of the subjects which should be paying attention to in the pylon's design process.

The corona performance of an overhead line is an important issue and depends on the electric field on the surface of line's conductors, which causes ionization of air adjacent conductors. If the conductor's surface gradients become higher enough, this will lead to facilitate self-sustaining ionization and thus, the corona discharge will occur near the surface of conductors [1]. The corona discharges on the surface of conductors are the source of macroscopic phenomena in the overhead lines, which can be characterized into Corona Losses, Radio Noise (RN) and Audible Noise (AN). Corona loss is the amount of energy, which is expended in the process of ionization, charge movement and recombination collisions around conductors and therefore, decreases the efficiency of power transmission. Another effect of corona discharges is the acoustic noise which its emission covers a wide range of audible spectrum. Annoyance sound caused by corona discharges should be considered by overhead line designers in order to get public acceptance by limiting the noise source.

Electron movement around conductor's surfaces is another consequence of corona discharge, which induces current pulses along the conductors. The pulse trains generate electromagnetic emissions over a wide frequency range in the vicinity of

T. Jahangiri et al., *Electrical Design of a 400 kV Composite Tower*, Lecture Notes in Electrical Engineering 557,
https://doi.org/10.1007/978-3-030-17843-7_8

overhead line (radiated emissions) [1] which cause interference to radio broadcast reception nearby the line. The accurate prediction of Radio Interference (RI) and Audible Noise levels are important factors at a design standpoint. The calculation of the levels of audible noise is simpler than the computation of radio noise levels. Because, audible noise is related only to the sound pressure waves generation and propagation, in which the sound propagation can be evaluated using the acoustics lows. The audible noise generation can be determined based on empirical formulas, generally as a function of the characteristics of conductor such as the diameter of conductor, its surface gradient, and the number of conductors in each phase bundle and weather condition. Unlike Audible Noise, the radio interference levels of an overhead line can be determined based on both semi-analytical and empirical RI prediction methods. Radio Interference (RI) depends on many parameters such as overhead line's geometry, phase conductor surface gradient and ambient conditions. The surface gradient on phase conductors has a direct impact on the generation of acoustic noise, which is presented in next section.

8.2 Surface Gradient on Phase Conductors

The surface gradient on the phase conductors is one of the parameters in empirical equations for the calculation of audible and radio noises as well as corona losses. Traditional empirical formulas employ a smooth conductor shape for phase conductors to evaluate the noise levels on the overhead transmission lines [2]. There are different approaches for computing electric field strength on phase conductors, which are based on analytical methods and numerical methods. The analytical methods have a restriction in application, because they cannot be deployed to simulate complex conductor surface geometries. In this study, a circular shape is adopted for the surface of phase conductors and finite element method is used to calculate electric field stress on the surface of phase conductors. Finite element method has the advantage of being able to compute electric field stresses for the arbitrary geometry of overhead lines and boundary shapes [2].

Electric field stress on the surface of phase conductors mainly depends on conductor voltage, its diameter, bundle and phase separation, and ground clearance. However, there are some other factors that affect the conductor surface stress along the overhead lines including: conductor sag, proximity of towers, uneven ground surface, and conductor stranding and protrusions [3, 4].

Due to the sag of conductors, the conductors will be closer to the earth surface and therefore increase the surface gradient on phase conductors compared at the tower top. This will lead to a non-uniform generation of audible noise and radio interference generation along the span. Therefore, an equivalent conductor height is defined in [1] with which height, a line with parallel conductors towards the ground plane would produce the same audible and radio noise levels as the actual line. For simplification, all above-mentioned factors are ignored and a simplified overhead line model can be created which consists of a number of cylindrical conductors with infinite length and parallel to each other and located above a flat earth surface, thus, three-dimensional overhead line can be represented by a two-dimensional model [4].

In order to calculate electric field stress on the surface of phase conductors, a 2D modeling of overhead line is built in ANSYS software. Martin ACSR conductors with a diameter of 3.617 cm are supposed to be used in the modeling of phase conductors in the form of a twin bundle with 40 cm bundle spacing. Shield wires are also considered in the modeling of the line and have a diameter smaller than phase conductors. As it can be seen in Fig. 8.1, the phase conductors are numbered and an asymmetrical voltage loading is applied to the double circuit overhead transmission line. The voltages applied on the phase conductors are based on the normal operation of the line (i.e. phase-to-phase voltage of 400 kV_{rms}).

Figure 8.2 shows the electric field distribution around the overhead line's conductors. From Fig. 8.2, it can be seen that conductor 2 has higher electric field stress with respect to other phase conductors because lower phase conductors are closer to the ground surface and therefore have higher electric field stress in comparison to middle and top phase conductors. Figure 8.2 shows that electric field distribution on the surface of conductor 2 is not uniform which is due to the proximity effect in the bundle however, it has a maximum value of 19.18 kV_{rms}/cm specified by a red color on its surface. The electric field magnitude on the surface of conductor 1 is slightly less than conductor 2. This diversity comes from the differences in the distances between both conductors (1 and 2) and other phase conductors.

In order to better understanding electric field distribution on the surface of phase conductors, electric field magnitudes around the surface of sub-conductors 1–6 are shown in Fig. 8.3 in which horizontal abscissa is related to the length of circumference of conductors.

Figure 8.3 shows that electric field distribution on the surface of phase conductors are in the form of sinusoidal shape with offset and conductor 2 has higher amplitude with respect to other conductors. Figure 8.3 also shows that the electric field stress profiles on the surface of two sub-conductors 1 and 2 (3 and 4 as well as 5 and 6)

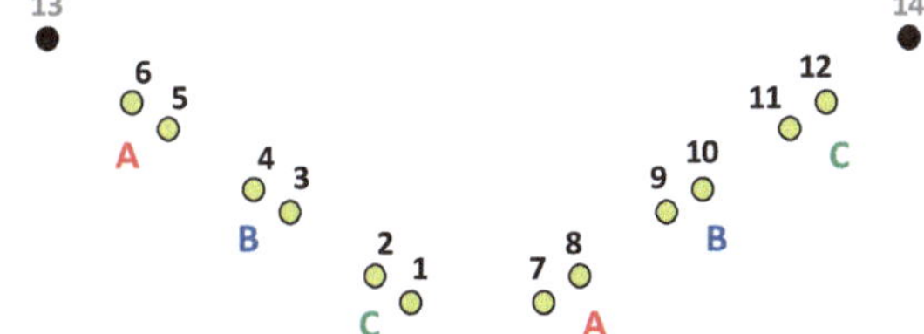

Fig. 8.1 Phase conductor arrangement and numbering of phase conductors and shield wires

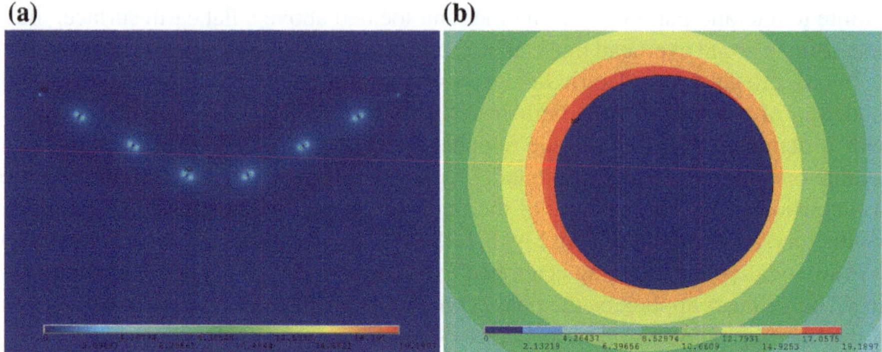

Fig. 8.2 **a** Electric field distribution around overhead line's conductors, **b** Electric field distribution around conductor number 2

within the same bundle are not the same as their conductor orientations produce two different curves with 180 degrees difference in the spatial distribution. The small change in the maximum amplitude of sub-conductor 1 and 2 is due to the effect of other phase conductors at the outside of the bundle. In Fig. 8.3, it is also evident that middle and top phase conductors have lower electric filed amplitudes because of the larger distance from the ground surface.

Table 8.1 gives the maximum electric field stress generated on the circumference of each sub-conductor. The maximum and average values of maximum electric field stress on the surface of sub-conductors are given in Table 8.1. In Table 8.1, 'maximum-maximum' implies to the maximum electric field magnitude among the 12 sub-conductors whereas 'average-maximum' implies to the mean value of maximum electric field magnitudes on the surface of 12 sub-conductors. Similarly, in

Fig. 8.3 Spatial distribution of electric field on the circumference of phase conductors

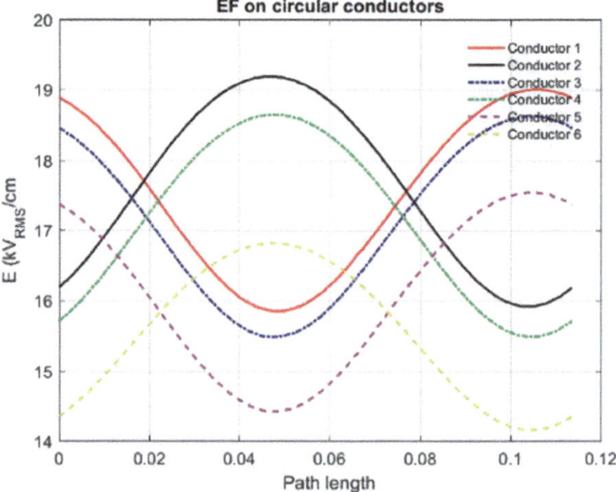

Table 8.1 Maximum and average values of electric field stress on the surface of sub-conductors

Conductor no	Circular conductor	
	EF_{max} (kV_{rms}/cm)	EF_{avg} (kV_{rms}/cm)
1	19.00	17.49
2	19.19	17.62
3	18.62	17.11
4	18.64	17.13
5	17.53	16.03
6	16.81	15.54
7	19.00	17.49
8	19.19	17.62
9	18.62	17.11
10	18.64	17.13
11	17.53	16.03
12	16.81	15.54
Average	'Average-maximum'	'Average-average'
	18.29	16.82
Maximum	'Maximum-maximum'	'Maximum-average'
	19.19	17.62

Table 8.1, firstly the average value of electric field magnitude on the surface of each sub-conductor is taken and afterward 'maximum-average' is chosen from the average electric field magnitudes on the surface of 12 sub-conductors. Subsequently, 'average- average' electric field magnitude among the 12 sub-conductors can be calculated by the arithmetic mean of the average electric field values. It should be mentioned that the electric field data for conductors 1–6 in Table 8.1 is repeated for the conductors 7–12, which is due to the symmetrical geometry of double circuit overhead transmission line. Table 8.1 shows that conductor 2 and 8 have the same maximum magnitude and therefore have higher average electric field magnitudes in comparison to other phases. The lowest electric field magnitude is also related to the conductors 6 and 12 at the top of the line.

In this research, it is assumed that the surfaces of conductors are clean and dry and corona discharge is initiated when the local electric filed stress exceeds 30 kV_{peak}/cm (peak value) [4], which is equivalent to 21 kV_{rms}/cm (RMS value). The electric field stress of 21 kV_{rms}/cm reflects the critical condition for the overhead transmission lines in service. For this reason, the electric field stress on the surface of phase conductors varies from 13 to 19 kV_{rms}/cm for the different overhead line designs [3]. In Table 8.1, it can be seen that the maximum electric field magnitude on the surface of conductors is 19.19 kV_{rms}/cm, which is approximately close to the threshold value of 19 kV_{rms}/cm used in overhead line designs. Therefore, maximum electric field magnitudes generated on the surfaces of phase conductors are in acceptable range.

The importance of maximum and average surface gradient is due to this fact that the electric field stress on the surface of conductors is a critical factor in the calculation of audible noise, radio interference and corona loss and it reflects the severity of these undesirable effects. The audible noise and radio interference level of an overhead line can be calculated based on the maximum values of electric field magnitudes on the surface of phase conductors whereas the corona power loss calculation depends on the average surface gradient on the phase conductors [4, 5]. In the next sections, the calculated maximum and average values of surface gradients in Table 8.1 are used to derive the audible noise, radio noise and corona loss of an overhead line consist of fully composite pylons.

8.3 Audible Noise

Audible noise is one of the major effects of power overhead lines. Audible noise has two different components. The first component is broadband crackling and hissing component, which is due to the energy dissipation from ionization process around conductors and therefore cause a sound emission to the surrounding the overhead line. The broadband component of audible noise has a considerable high-frequency spectrum that distinguishes it from the environmental noises.

The second component of audible noise is the hum component which is created by the periodically movement of the positive and negative ions around phase conductors. By the movement of ions, the energy transfers to the surrounding air molecules and causes humming noise. The hum component of audible noise has low-frequency content which is equal to twice the power frequency. The frequency of hum noise is 100 Hz for a 50 Hz AC system. The human ear has a different response to the two frequency components and is more sensitive to high-frequency component which is in the range of 1–20 kHz [3]. Therefore, A-weighting filter is used to measure the human response and consequently, the audible noise level is expressed as dBA above 20 μPa. In comparison to A-weighted broadband component, the hum component has less annoying effect in the surrounding area. For this reason, the A-weighted broadband level is only determined in industrial noise regulations [6]. However, although audible noise is one of the design factors in overhead line's planning and design, there is no specific regulation regarding the limit of audible noise level that an overhead transmission line may produce. Generally, the design of overhead lines are based on preventing corona discharges under dry condition and audible noise generation under fair climate conditions is usually negligible in comparison to the foul weather condition [3]. Since foul weather condition is the worst case for audible noise, therefore, the design criterion for overhead lines is fixed on the audible noise performance of the line under wet conditions [3]. The mechanisms of audible noise generated by overhead lines are still debatable [3], therefore, only empiric formulas are widely used to compute the audible noise level. The empirical formulas are driven from laboratory cage testing or full-scale field measurements of audible noise during rain [7]. The A-weighted audible noise levels can be determined by two widely

Table 8.2 Average-maximum surface gradient in each bundle of overhead line

Phase	Bundled conductors	'Average-maximum' bundle gradient
		Avg. EF_{max} (kV_{rms}/cm)
C	Sub-conductor 1 and 2	19.01
B	Sub-conductor 3 and 4	18.63
A	Sub-conductor 5 and 6	17.17
A	Sub-conductor 7 and 8	19.01
B	Sub-conductor 9 and 10	18.63
C	Sub-conductor 11 and 12	17.17

employed methods of EPRI (Electric Power Research Institute) method and BPA (Bonneville Power Administration) method [8].

8.3.1 *Audible Noise Results and Discussions*

The acoustic performance of the overhead line with fully composite pylons is presented in this section for different weather conditions. Electric field stress on the surface of phase conductors is one of the key factors in the calculation of generated audible noise by the overhead line. According to Table 8.1, average-maximum surface gradient in each bundle of the overhead line are given in Table 8.2 which are used to calculate audible noise level of each phase (A, B, and C). The conductor numbering and phase arrangements are based on Fig. 8.1.

As it can be seen from Table 8.2, the highest electric field magnitude is related to lower phase conductors whereas the lowest electric field magnitude is related to top phase conductors. Therefore, lower phase conductors will produce higher audible noise in comparison to top phase conductors. However, the contribution of all phase conductors in audible noise generation gives the audible noise level of the line. The shield wires do not significantly contribute to the total audible noise of the overhead transmission line because they have small size, small numbers and very low electric field stress on their surfaces [9]. For this reason, the effect of shield wires has not been considered in the empirical formulas.

In the evaluation of audible noise of overhead lines, it is recommended in IEEE standard 656-1992 [9, 10] that the calculation point should be considered at 1.5 m above ground and 15 m measured horizontally from the outside phase conductor of the line as shown in Fig. 8.4. It is noted that the value obtained at this location gives the audible performance of the overhead lines. In order to get the sound pressure levels at 1.5 m above ground and 15 m horizontally from the outside phase, the lateral profile of audible noise based on the EPRI and BPA methods are calculated and presented for different weather conditions.

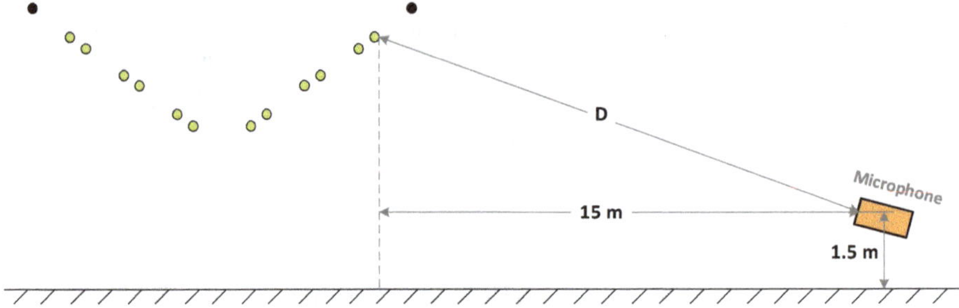

Fig. 8.4 Calculation point in the evaluation of audible noise level of overhead line

Fig. 8.5 Contribution of phase's noise and resultant audible noise for heavy rain (EPRI)

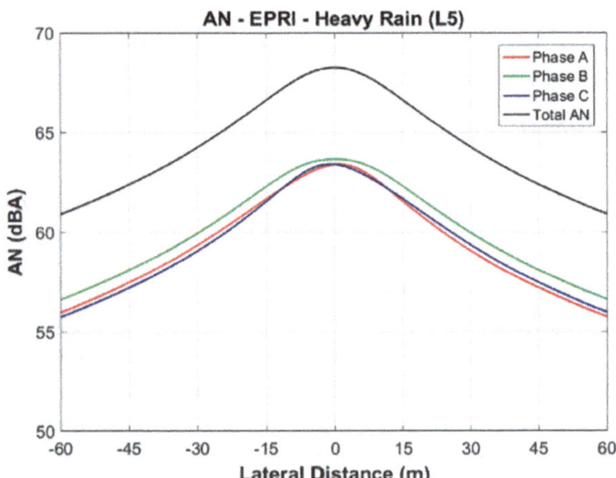

8.3.1.1 Lateral Profile of Audible Noise Based on EPRI Method

The generated sound pressure level during heavy rain (L_5 level) is calculated based on EPRI method, and illustrated in Fig. 8.5. In Fig. 8.5, the produced audible noise by different phases and the resultant audible noise level are shown for the asymmetrical phase arrangement on the overhead line. The total audible noise level of the line is obtained by the anti-logarithmically summation of the contribution of each phase's noise.

Figure 8.5 shows that the conductors of phase A and C produce the same audible noises with a small horizontal shifting along their lateral profiles. This is due to the asymmetrical arrangement of phase conductors in the line as shown in Fig. 8.1 which causes that the average electric field stress of both phases of A (or C) becomes lower than the average electric field stress of both phases B. This leads to a higher audible noise generation by phase B in comparison to other two phases of A and C (shown in Fig. 8.5). Moreover, Fig. 8.5 shows that the total audible noise produced by all phases has a peak value of 68.26 dBA at the center of the line.

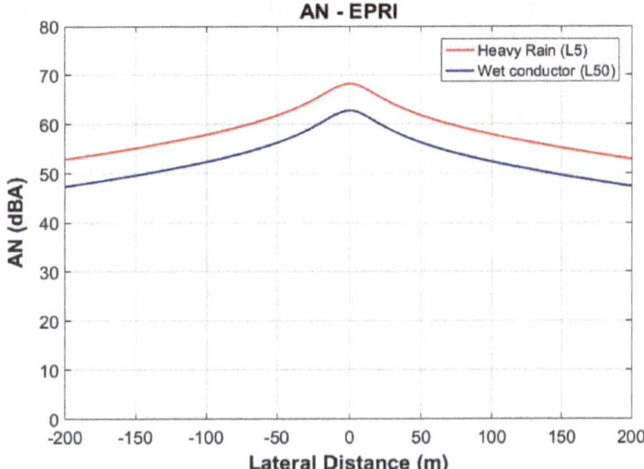

Fig. 8.6 Lateral profile of audible noise for different weather conditions (EPRI)

The lateral profiles of audible noise for different weather conditions of heavy rain (L_5) and wet conductor (L_{50}) are shown in Fig. 8.6 in which the total sound pressure level of overhead line is presented. The EPRI method does not provide a formula for the calculation of audible noise during fair weather condition. However, Fig. 8.6 shows that heavy rain condition produces higher audible noise level in comparison to the wet conductor condition. The difference between the two weather conditions is 5.46 dBA at the center of the line.

8.3.1.2 Lateral Profile of Audible Noise Based on BPA Method

The lateral profiles of sound pressure level during heavy rain (L_5 level), wet conductor (L_{50} level) and fair weather (L_{50d} level) are displayed in Fig. 8.7. Figure 8.7 implies the fact that the overhead transmission line produces maximum amplitudes of audible noise during heavy rain condition. Moreover, it shows that the amplitudes of audible noise during fair weather condition is much lower than the other weather conditions.

The public annoyance regarding audible noise during heavy rain (L_5 level) may not be a concern because the background sound produced by the rain falling and wind covers the audible noise of the overhead line [7]. On the other hand, people prefer to stay inside the building in heavy rain condition and therefore the ones at the outside of building do not expect to have a silent condition and thus do not feels inconvenience due to audible noise. Therefore, the level of audible noise during foul and fair weathers should be paying attention to the evaluation of overhead line's audible noises.

8.3.1.3 Comparison of Audible Noises Based on EPRI and BPA Methods

The profiles of audible noise based on EPRI and BPA methods are displayed in Fig. 8.8. The comparison of audible noise results during heavy rain shows that EPRI method predicts higher audible noise levels for the overhead line. The difference between two methods reaches to 2.7 dBA at the center of the line. Furthermore, it can be seen that during wet conductor condition, there is an approximately good agreement between the results of audible noises by both methods.

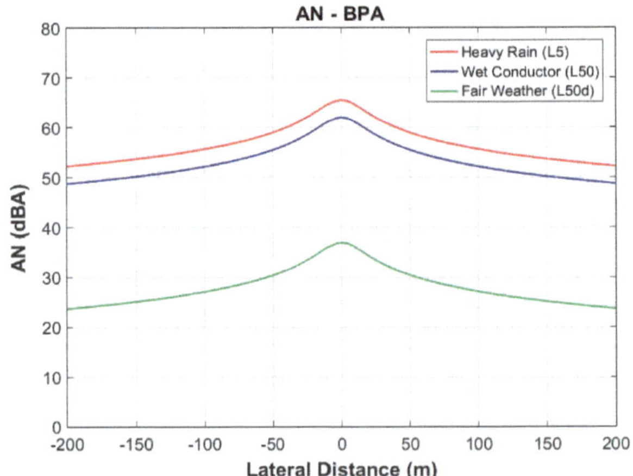

Fig. 8.7 Lateral profile of audible noise for different weather conditions (BPA)

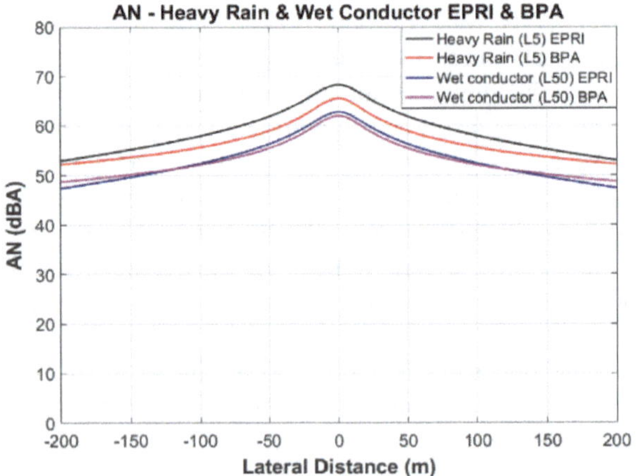

Fig. 8.8 Comparison of audible noise levels based on EPRI and BPA

Table 8.3 Audible noise level at 1.5 m above ground and 15 m from the outside phase

AN method	Heavy rain (L_5) (dBA)	Wet conductor (L_{50}) (dBA)	Fair weather (L_{50d})
EPRI	65.10	59.58	–
BPA	62.15	58.65	33.65 dBA

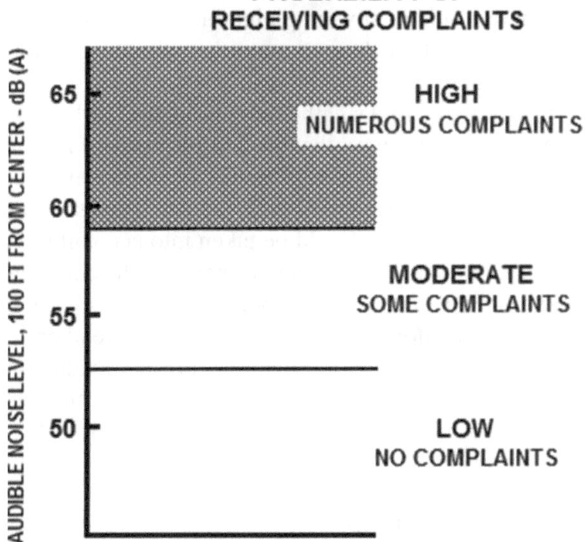

Fig. 8.9 Expected public response from overhead line audible noise levels [15]

8.3.2 *Acoustic Performance of an Overhead Line Composed of Fully Composite Pylons*

The audible noise levels at a specific location near the overhead line are given in Table 8.3 for the different weather conditions and prediction methods. The calculated values are related to audible noise level at 1.5 m above ground and 15 m measured horizontally from the outside phase of the overhead line.

So far, any specific standard has not been established as the control criteria for interpreting the generated audible noise from overhead transmission lines [5]. However, there are some guidelines in the literatures [5, 11, 12, 13] to evaluate the acoustic performance of the lines. The probability of receiving complaints (shown in Fig. 8.9) presented by Perry is a widely used rule for the evaluation of acceptability of audible noise levels under overhead lines [14, 15]. This probability is presented for foul weather condition (L_{50} wet conductor).

However, there are design criteria for the acoustic performance of AC lines which are determined based on subjective evaluation obtained from a group of people as follows [12]:

- Low complaints: AN < 52 dBA
- Moderate (some) complaints: 52 < AN < 58 dBA
- Many complaints: AN > 58 dBA

According to Table 8.3, audible noise level during heavy rain conditions exceeds 58 dBA, however, as it is mentioned before, audible noise during heavy rain condition might not be a concern in the design of overhead lines. On the other hand, the audible noise during fair weather is 33.65 dBA which is much less than the 52 dBA. It means that there would be no complaint due to audible noise during fair weather condition and therefore it has less of importance because of its low level at both sides of the overhead line.

In the case of foul weather condition, the IEEE Task Force of the Corona and Field Effects (1982) reported that the audible noise of AC overhead lines causes a concern only in the foul weather condition [14]. Moreover, international experiences with AC overhead transmission lines have indicated that the audible noise during foul weather (L_{50} wet conductor) should be taken into account to avoid public complaints [11].

Based on Table 8.3, it can be seen that the calculated audible noise level by both methods (EPRI and BPA) is about 59 dBA which is in the range of many complaints. Therefore, the noise generation during foul weather condition, as a criterion in overhead line design, is higher than the threshold value of 52 dBA and it seems that the generated audible noise does not comply with the requirement for residential areas in the vicinity of the overhead line. According to Fig. 8.8, the audible noise of the line attenuated to 52 dBA at the distance of 100 m from the center of the line and therefore, at 100 m away from the line there would be low complaints about the audible noise of the line.

In order to better compare audible noise produced by the overhead line composed of fully composite pylons with other traditional overhead lines, the audible noise generated by an overhead line composed of L_2 towers is calculated using BPA method and displayed in Fig. 8.10 for foul weather condition. It can be seen that the audible noise generated by both overhead lines (composed of Fully composite pylons and L_2 towers) differ slightly and the difference between them at 15 m horizontally from the outside phases is about 3.5 dBA.

A slightly higher audible noise level of fully composite pylon is due to the geometrical configuration of fully composite pylon which differs from L_2 tower. Increasing the number of sub-conductors in a bundle or increasing conductor's diameter will increase the generated audible noises from the fully composite pylon. The only parameter, which can effectively reduce the audible noise level of fully composite pylon, is increasing the height of fully composite pylon.

8.4 Radio Noise

One of the important factors in electromagnetic compatibility of high voltage overhead transmission lines is radio noise produced by corona discharges around phase

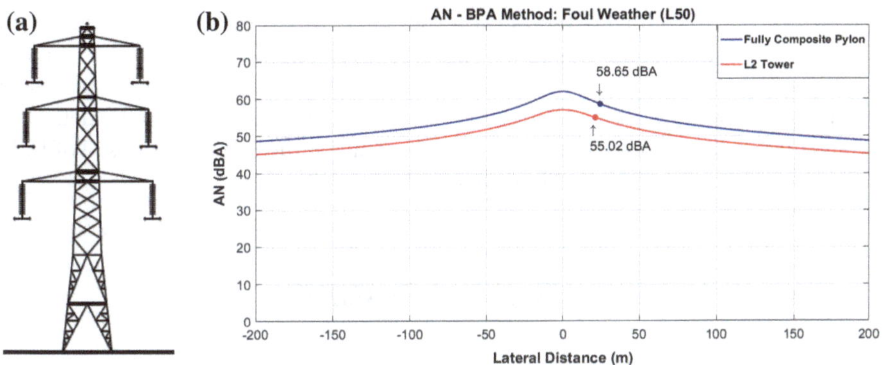

Fig. 8.10 **a** L_2 tower, **b** Audible noise produced by L_2 tower and fully composite pylon

conductors. The radio noise caused by corona discharge has a frequency between 3 kHz and 30 MHz [16]. In electric utility industry, radio noise is the radio interference to the AM broadcast band which covers the frequency range of 535–1605 kHz [9]. Radio interference in the AM broadcast band is severe in foul weather condition because during foul weather condition, corona activity on the surface of conductors is much higher than in dry weather. However, there are a few complaints to radio interference for radio broadcast band in last decade. This is mainly due to the popularity of FM broadcast band which is not affected by overhead line radio interference [9].

Radio interference level of overhead transmission lines is dependent on many parameters such as geometrical characteristics of overhead transmission line, surface gradient on phase conductors, and climatic conditions. A lot of researches on radio interference have been conducted and several methods to estimate radio interference level have been developed. There are two basic radio interference predicting methods. One is empirical methods such as CIGRE (France), BPA (USA) which are formulated based on masses of experimental data whereas another is semi-analytical methods based on excitation functions such as EdF (France), IREQ (Canada) and EPRI (USA) methods [1]. Radio interference cannot be predicted entirely based on analytical methods because it is not practical due to the large number of affecting parameters and complex process of corona discharges. A large amount of database of radio noise levels have been recorded for various overhead line's geometries and climate conditions and has been analyzed to produce empirical formulas for predicting radio interference of a proposed overhead line.

Radio noise generation can be decomposed into two quantities; geometrical quantity and generation density. Generation density is the excitation function that is independent from the geometry and considers random and pulsative nature of induced currents in phase conductors which can be measured in corona cage. The measured excitation function can be used to compute radio noise of overhead line by considering it in the analytical propagation model [1] and therefore, this semi-analytical method can better estimates the radio noise levels of overhead line designs with different geometries (in comparison to the empirical methods which makes it theoretically and practically significant). The procedure and formulas of semi-analytical and empirical methods for prediction of radio noise are presented in more details in [8], and the predicted radio interference levels of an overhead line composed of fully composite pylons are presented in next section.

8.4.1 Radio Noise Results and Discussions

The prediction of radio interference levels based on the empirical and semi-analytical methods is presented in this section for different weather conditions. One of the parameters, which is needed to calculate the radio noise level of an overhead line, is the electric field stress on the surface of phase conductors. The electric field magnitudes in Table 8.2 are used to predict the radio noise performance of the overhead line composed of fully composite pylons. According to Fig. 8.1, asymmetrical phase arrangements are assumed to be applied to the line.

In order to represent the contribution of each phase in total radio noise level of overhead line, the predicted radio noise field by CIGRE and EdF methods are illustrated in Fig. 8.11 for average fair weather condition (L_{50d}). Figure 8.11a and b show the contribution of different phases (A, B and C) for calculating total radio noise of the line. Radio noises of all phases in Fig. 8.11a have different curve shape rather than the noises shown in Fig. 8.11b which is due to the difference in radio noise prediction methodologies. Predicted noise levels in Fig. 8.11a are computed by using an excitation function approach (EdF) whereas the noise levels in Fig. 8.11b are driven from an empirical formula (CIGRE). Figure 8.11 also shows that the EdF method estimates higher radio noise levels for the line in comparison to the CIGRE method. In the determination and evaluation of radio noise generated by overhead transmission lines, CISPR standard [17, 18] recommended that a reference point (shown in Fig. 8.12) should be considered at 2 m above ground and 15 m laterally from the outermost phase conductor. It is mentioned that the value obtained at this point gives the radio noise level of the line.

The calculations of radio noise in other distances than 15 m are required to get the lateral profile of radio noise at the vicinity of the line. This lateral profile is useful to estimate the width of the corridor along the pathway of overhead line [18] because sometimes the right-of-way of an overhead line may be determined by radio noise criteria [13].

In order to obtain the radio interference level at the height of 2 m above ground and 15 m horizontally from the outside phase, the lateral profiles of radio noise are calculated based on the empirical and semi-analytical methods and are presented in next section for different weather conditions.

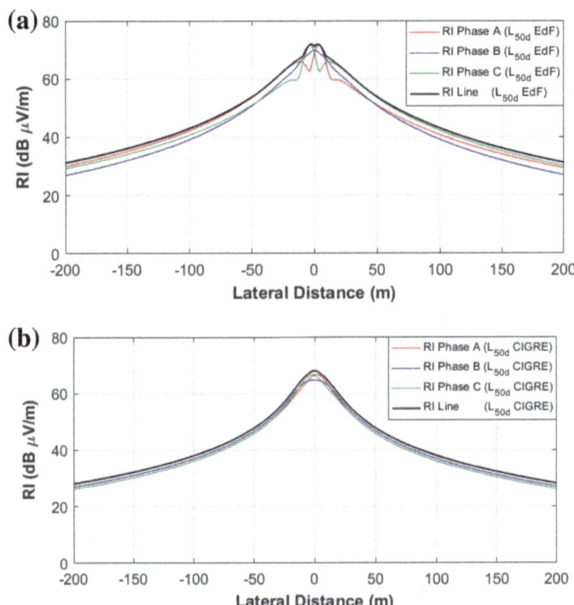

Fig. 8.11 Contribution of phases in total radio noise level of overhead line based on **a** EdF method, **b** CIGRE method for average fair weather condition (L_{50d})

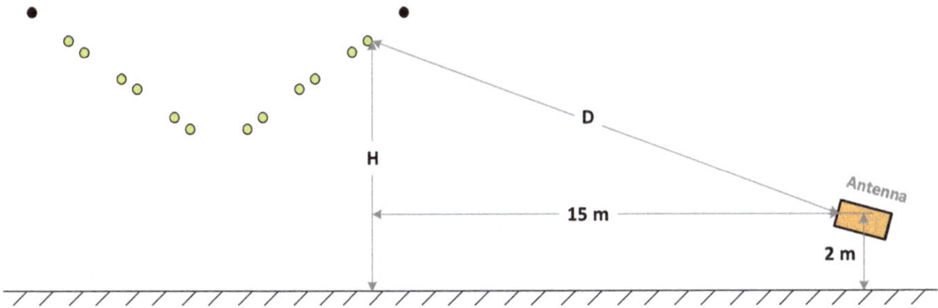

Fig. 8.12 Antenna position at 2 m above ground and 15 m laterally from the outermost phase conductor

8.4.1.1 Lateral Profiles of Radio Noise

In this section, lateral profiles of radio noise are presented for different weather conditions, which are as follows:

- L_{1w}: Heavy rain (maximum foul)
- L_{5w}: Rain (stable foul)
- L_{50w}: Average foul
- L_{50d}: Average fair

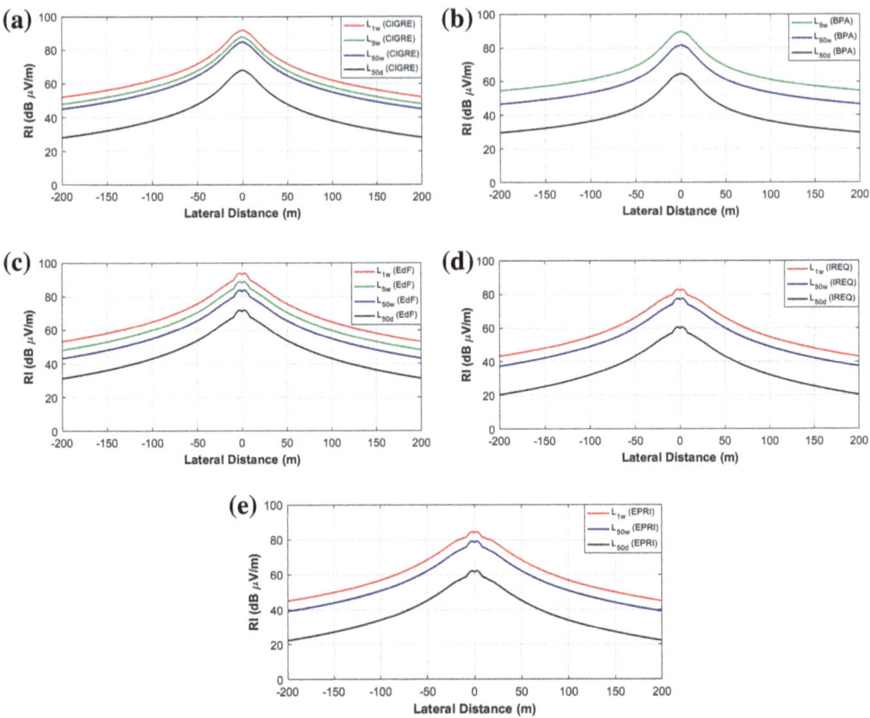

Fig. 8.13 The predicted lateral profiles of radio noises based on empirical methods of **a** CIGRE, **b** BPA—and based on semi-analytical methods of **c** EdF, **d** IREQ and **e** EPRI for different weather conditions

The predicted lateral profiles of radio noise based on the empirical methods of CIGRE and BPA and semi-analytical methods of EdF, IREQ and EPRI are displayed in Fig. 8.13 for different weather conditions. Figure 8.13 shows that the radio interference levels of the line depend on the weather condition and the radio noise prediction method. The lateral profiles based on different methods shown in Fig. 8.13 implies that the heavy rain condition (L_{1w}) has maximum amplitudes with respect to other weather conditions. This comes from the fact that corona activity and electric field stress on the surface of wet conductors is very high during heavy rain condition whereas dry conductors in fair weather condition generate lower electric field stress. This leads to the prediction of radio noise with low amplitudes for the average fair weather condition (L_{50d}). However, the predicted noise by EdF and CIGRE gives a higher values in comparison to other methods and the noise levels calculated by IREQ and EPRI have lower values. It should be mentioned that BPA method does not provide a formula for the calculation of the radio noise during heavy rain (L_{1w}). The same condition is for the IREQ and EPRI methods which do not have any estimation for the stable foul (L_{5w}).

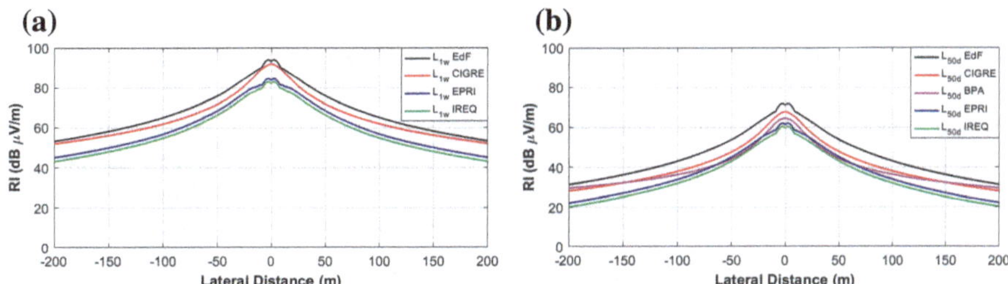

Fig. 8.14 Radio noise level for weather category of a) heavy rain (L_{1w}) and b) fair weather (L_{50d}) based on different noise prediction methods

For the weather category of heavy rain (L_{1w}) and fair weather (L_{50d}), Fig. 8.14 indicates that there is a good agreement between the radio noise prediction of two methods of IREQ and EPRI. Moreover, the two methods of EdF and CIGRE give a higher estimation for the generated radio noise by the line.

8.4.2 Radio Noise Performance of Line

According to [9], the radio interference level of the line is determined based on the frequency of 0.5 MHz, CISPR type receiver and a distance of 15 m from the outside phase. IEC/CISPR standard has tried to define allowable radio interference limits for overhead transmission lines, however, this could not be done because many nations prefer to define national limits for the radio noise level from overhead lines. As an example, the Swiss standard (1966) stated that the radio interference level during fair weather condition should not exceed 34 dB for lines operating at less than 100 kV and 46 dB for lines operation at more than 100 kV [9].

On the other hand, it is reported in [19] that the recommended limits by IEEE (1971) and Industry Canada (2013) for the radio noise level at 15 m beyond the outermost conductor are 61 dB μ V/m and 60 dB μ V/m, respectively. These limits are defined for fair weather condition. In this study, the radio noise limit of 60 dB μ V/m in fair weather condition is adopted as a criterion to evaluate and interpret the radio noise results obtained from empirical and semi-analytical methods. The radio noise levels at the specific location of 2 m above ground and 15 m measured horizontally from the outermost phase of the line are given in Table 8.4 for different weather conditions and prediction methods.

From Table 8.4, in fair weather category, all prediction methods except EdF give an estimation of radio noise level below the criterion value of 60 dB μ V/m. The value obtained by EdF method is 2.73 dB μ V/m above the criterion value and the lowest predicted value of 52.28 dB μ V/m in fair weather category is related to IREQ method. Therefore, it can be said that the radio interference performance of the overhead line composed of fully composite pylons is in acceptable range. Table 8.4

Table 8.4 Radio noise levels at the specific location (2 m above ground and 15 m measured horizontally from the outermost phase) of the line for different weather conditions and prediction methods

Method	Average fair (L_{50d})	Average foul (L_{50w})	Stable foul/rain (L_{5w})	Maximum foul/heavy rain (L_{1w})
CIGRE	58.42	75.42	78.42	82.42
BPA	53.85	70.85	78.85	–
EdF	62.73	74.73	79.73	84.73
IREQ	52.28	69.28	–	75.50
EPRI	54.36	71.36	–	77.63

also shows that in stable foul condition, there is a good agreement between the calculated values by CIGRE, BPA and EdF methods which predict a radio noise level of approximately 79 dB μ V/m for the line. The maximum predicted values in Table 8.4 are related to the CIGRE and EdF methods for the weather category of heavy rain. It is mentioned in [20] that the radio interference level under heavy rain has maximum level in comparison to other weather conditions and this weather category should be introduced as determinant weather category in the determination of noise limit for any overhead transmission line.

8.5 Corona Loss

Power loss is one of the consequences of corona discharges around phase conductors which is known as corona loss. The power loss due to corona drawing energy from high voltage source connected to the overhead transmission line. One of the determinant factors in the economic choice of conductors is its corona loss which is dependent on the value and variation of corona losses during different climate conditions. However, in higher voltage overhead transmission lines which utilize bundled conductors, radio interference and audible noise have important role than corona loss in the choice of conductors [9]. In the design of overhead lines, the trend is to minimize corona loss in dry conditions (fair weather), however, the humid and rainy weather conditions aggravate the corona losses and increase the power loss up to several hundred of kW/km [21].

During the development of overhead transmission lines to higher voltages, corona loss in fair weather condition was one of the important design criteria for choosing phase conductors and the phase conductor's diameter was chosen based on not exceeding a specific level of corona loss in fair weather condition.

One of the first empiric formulas for the calculation of corona loss under dry and clean conductors was introduced by Peek [9]. Comparing corona loss computed by Peek's equation with experimental data has shown that Peek's formulae cannot be

Table 8.5 Acceptable limits for corona loss during fair weather condition [24]

Overhead line voltage (kV)	(kW/km) per phase
132	0.02
230	0.2
400	0.33

applied for the voltage levels near to corona onset which is the normal condition for operating practical overhead transmissions lines [9]. Another drawback of Peek's formulae is that it gives too high loss for small conductors and gives too low loss for large conductors [9]. Another empirical formulae for calculating corona loss developed by Peterson [9]. Experimental researches have shown that there is good agreement between the measured corona loss and calculated corona loss by Peterson's formulae [9], whereas the Peek's formula estimates much higher corona loss than the measured ones. However, the Peterson's formula is capable to calculate corona losses of overhead transmission lines with a single conductor in each phase and therefore can only be applied to estimate corona loss of overhead lines with the voltage levels of below 230 kV [9].

Other empirical formula were presented by Electricite de France (EdF) which can be used for the evaluating corona loss of overhead transmission lines with single or bundled conductors for both fair weather and foul weather conditions [9]. In comparison foul weather, corona losses during fair weather conditions are insignificant for the majority of overhead lines above 230 kV, however, corona loss during fair weather occur for a higher percentage of time and may affect the annual corona losses.

8.5.1 Calculated Corona Losses and Discussion

In the design of an overhead transmission line, corona loss in fair weather condition should be small enough to increase the efficiency of the line. A properly designed overhead line usually produces insignificant corona loss during fair weather condition [22]. Corona losses in extra high voltage overhead lines vary between a few kW/km in fair weather up to hundred kW/km during thunderstorm and bad weather conditions [23]. However, acceptable limits for corona loss during fair weather condition are given in Table 8.5 for different voltage levels [24].

In order to compare the calculated corona losses with the acceptable limits, the average corona losses for the overhead line composed of fully composite pylons are given in Table 8.6. From Table 8.6, it can be seen that the calculated corona loss by Peek's formula gives much higher losses with respect to the EdF and BPA formulas as well as the acceptable limit of 0.33 (kW/km/phase) for fair weather condition. It was mentioned before that Peek's formula has some drawbacks in corona loss estimations since it is not suitable for predicting corona losses of extra high voltage

Table 8.6 Estimated corona loss for overhead line composed of fully composite pylons

For one circuit of overhead line	Corona loss in fair weather			Corona loss in foul weather
	Peek	EdF	BPA	BPA
Average corona loss (kW/km) per phase	10.77	0.34	0.20	10.32

lines and it also gives too high loss for small conductors and gives too low loss for large conductors [9]. The calculated values by EdF is in good agreement with the acceptable level of 0.33 (kW/km/phase). The calculated value by BPA is lower than the acceptable limit of 0.33 (kW/km/phase) and predicts lower corona losses for the overhead line during fair weather condition. The corona loss performance of the line during foul weather condition is calculated by BPA formula and is about 10 kW/km per phase. The higher corona loss in comparison to fair weather losses is due to higher corona activities and electric field stresses on the surface of phase conductors in foul weather condition.

As a result, the calculated corona losses by EdF and BPA formulas satisfy the acceptable limit for fair weather condition and it can be said that the corona loss produced by the overhead line composed of fully composite pylons is in a reasonable range.

8.6 Electromagnetic Emissions

Electromagnetic emissions of an overhead line composed of fully composite pylons is one of the subjects which should be evaluated in the design process of the new pylon. The biological effects of overhead lines, as one of the consequences of power-frequency electromagnetic radiation, have been investigated in many literatures. Many researches have been done to measure and compute the electric and magnetic fields around power lines such as [25–28]. Some alternatives for reducing electric and magnetic fields are reported in [28–30]. In [31], two- and three-dimensional methods are used to compute power-frequency electromagnetic field produced by overhead lines. Finite element method is used in [32] to calculate electric and magnetic fields of overhead lines. In most cases of electromagnetic field calculations, overhead line conductors assumed to be parallel to the flat ground and the conductor's sag is ignored within the span length. Moreover, the average height of conductors above the ground surface has been adopted in electromagnetic field calculations [33]. However, in [34, 35] the effect of sagged conductors has been considered in the calculation of electric and magnetic fields around overhead lines. Electric and magnetic field distribution within the whole span is also investigated in [31, 32].

By increasing the overhead line's voltage level, the ground level of electric and magnetic fields of power lines increases. This raises the concerns for the people who live in the vicinity of high voltage and extra high voltage overhead lines. In this regard, all countries define corridors around overhead transmission lines as Right Of Way (ROW) width. In order to eliminate any risk for human health and its safety, adequate ROW width should be considered around the overhead lines by maintaining induced electric and magnetic fields near to the ground's surface within acceptable levels. In this regard, International Commission on Non-Ionizing Radiation Protection (ICNIRP) and World Health Organization (WHO) have determined some precaution values for electric and magnetic fields below overhead lines [13]. Based on the guidelines, emission of electric and magnetic fields around and under overhead lines at 1 m above ground surface should be kept below the precaution values.

In this section, ANSYS finite element software is used to compute electric and magnetic fields around an overhead line composed of fully composite pylons. Moreover, the results of finite element approach are compared with the results of analytical approaches to calculate electric and magnetic fields around the overhead line. Finally, the necessary right-of-way (ROW) width for the overhead line composed of fully composite pylons is derived from analytical and finite element results and determined based on the authentic standards.

8.6.1 Phase Conductor Arrangements

In double circuit overhead transmission lines, the arrangements of phase conductors have a significant effect on the generated electric and magnetic fields around and under the power lines [36]. Some researches in [25, 32, 36] showed that optimum phase arrangement is a suitable way to minimize the emission of electric and magnetic fields in comparison to symmetrical phase arrangement. The calculations in [25, 32, 36] prove that there is a drastic reduction in electric and magnetic field levels by applying the optimum phase conductor arrangement.

Figure 8.15a shows the application of symmetrical phase arrangement for the overhead line composed of fully composite pylons and Fig. 8.15b displays the optimum phase arrangement on the pylon. By applying two different phase conductor arrangements, electric and magnetic field distributions at one meter above the earth surface (under the overhead line) are calculated and compared by using analytical and finite element methods. The electric field strength of E is used to represent the electrical field around the power lines. Instead of the magnetic field strength of H, the magnetic flux density of B is often used for determining magnetic field around the power lines because it can be measured easily [13].

Analytical approach for the calculation of electric field level around the overhead transmission lines is presented in [8]. It should be mentioned that this method does not take into account the effect of shield wires in the calculations. And only the resultant electric field produced by phase conductors can be calculated at one meter above the earth surface. As it will be shown, finite element method has an advantage and can be used to consider the effect of shield wires on the generated electric field around the overhead line. Similarly, there are analytical formulae for computing magnetic flux density at one meter above the earth surface which are presented in

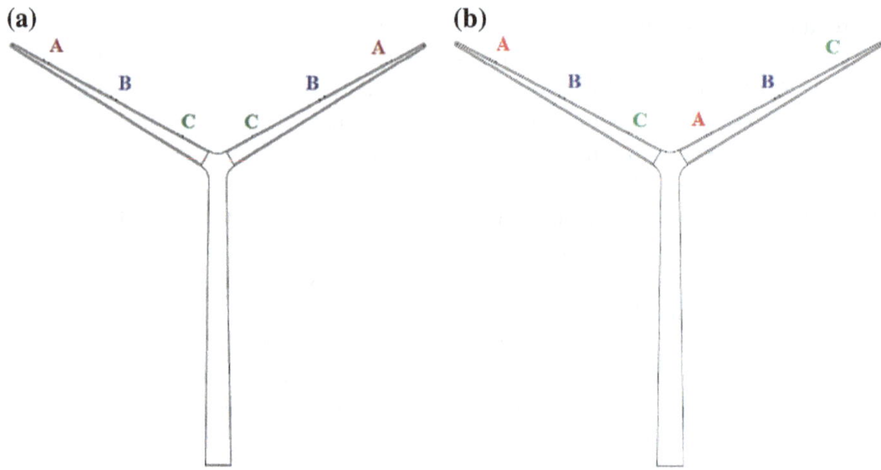

Fig. 8.15 **a** Symmetrical phase arrangement, **b** Optimum phase arrangement

[8]. Since the shield wires do not carry current flowing, therefore, their contribution on the produced resultant magnetic flux density can be ignored. Thus, analytical method and finite element method should provide the same results for the magnetic field distribution.

8.6.2 Analytical and Finite Element Method Results and Comparison

8.6.2.1 Electric Field (E)

Figure 8.16 shows the electric field magnitudes at one meter above the earth surface which is calculated based on FEM and analytical formulae for the overhead line with symmetrical phase arrangement. As it can be seen, the results of finite element modeling coincide with the analytical formulae for the overhead line without shield wires. Electric field profile calculated by FEM for the overhead line with shield wires is also depicted in Fig. 8.16. In comparison to the case without shield wires, the presence of shield wires (in electric field calculations) increases electric field magnitudes under the overhead line and reduces electric fields in both sides of the line.

Similarly, the electric field magnitudes at one meter above the earth surface are illustrated in Fig. 8.17 for optimum phase conductor arrangement. From Fig. 8.17, it can be seen that electric field magnitudes in the calculations with shield wires are lower than the calculations without shield wires. Figures 8.16 and 8.17 imply that there is good agreement between the results of FEM with analytical formulae in the

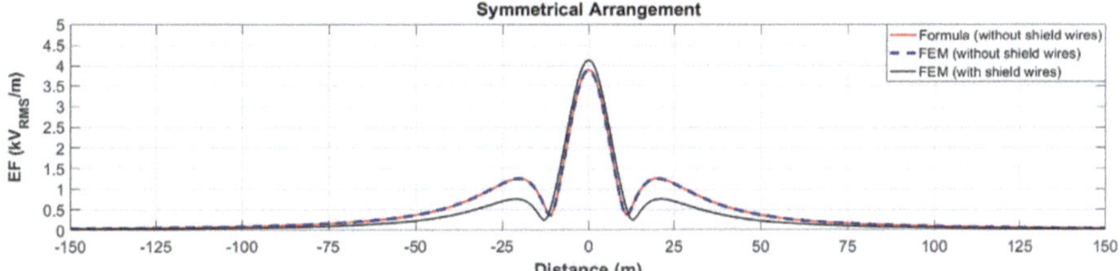

Fig. 8.16 Electric field magnitudes at one meter above the earth surface based on FEM and analytical formulae for the overhead line with symmetrical phase arrangement

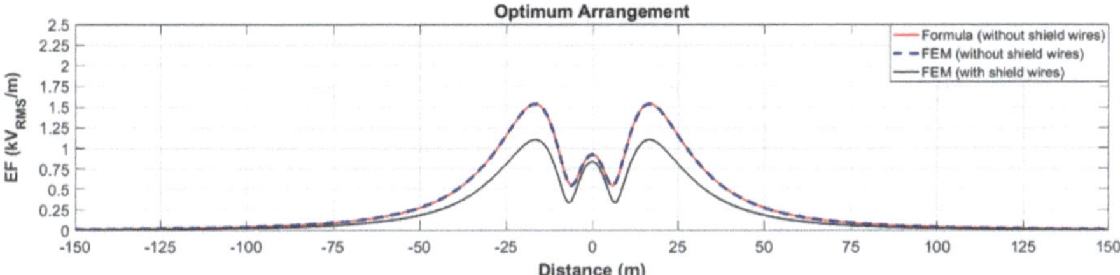

Fig. 8.17 Electric field magnitudes at one meter above the earth surface for optimum phase conductor arrangement

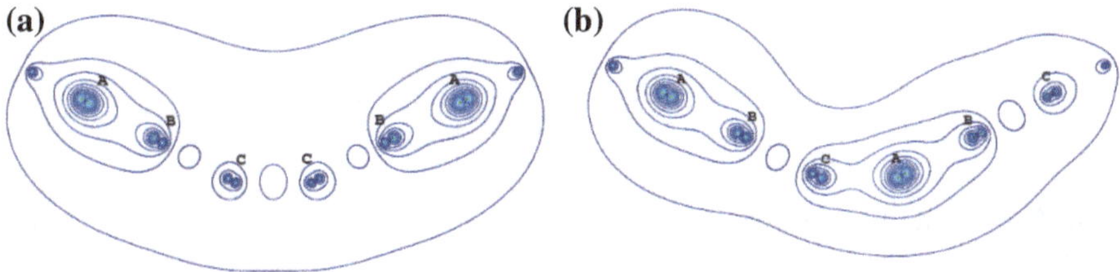

Fig. 8.18 Electric field distribution lines based on phase conductor arrangements of **a** symmetrical and **b** optimum

calculation of electric field magnitudes (without shield wires) at one meter above the earth surface.

Electric field distribution lines around the overhead line composed of fully composite pylons are depicted in Fig. 8.18 for the two phase conductor arrangements of symmetrical and optimum. Electric field magnitudes at one meter above the earth surface are shown in Fig. 8.19 for the two phase arrangements of symmetrical and optimum. It is evident that the optimum phase conductor arrangement produces much less electric field magnitudes with respect to the symmetrical arrangement.

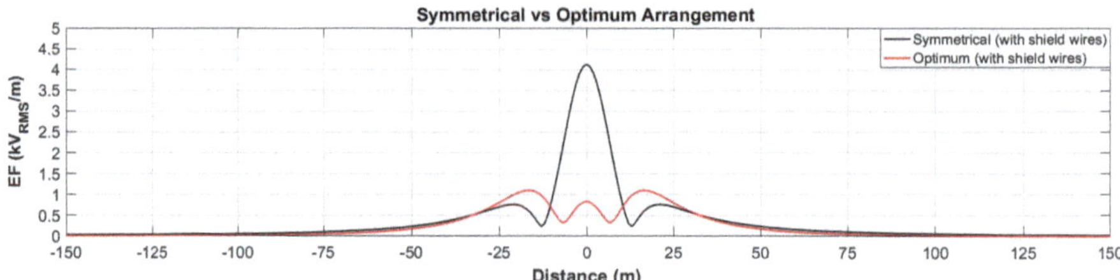

Fig. 8.19 Electric field magnitudes at one meter above the earth surface based on phase arrangements of symmetrical and optimum

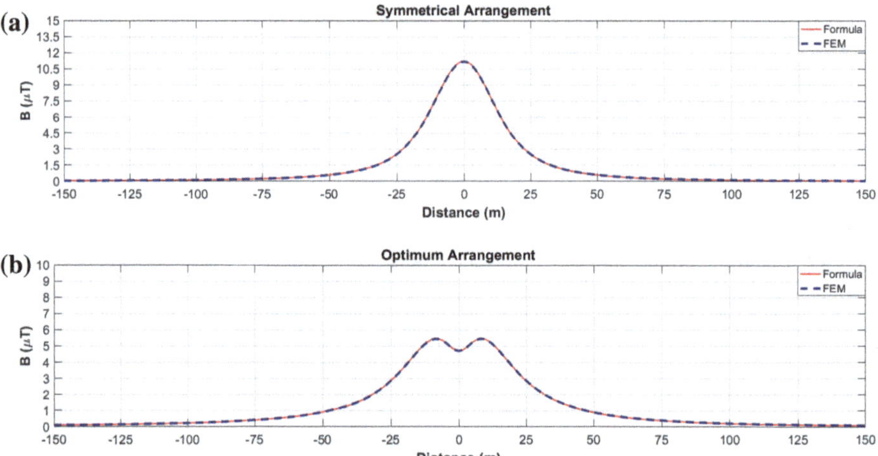

Fig. 8.20 Magnetic flux density at one meter above the earth surface for **a** symmetrical and **b** optimum phase arrangements

8.6.2.2 Magnetic Field (B)

Magnetic flux density at one meter above the earth surface is shown in Fig. 8.20 for the symmetrical and optimum phase arrangements. According to Fig. 8.20a and b, magnetic flux densities calculated by FEM and analytical formulae coincide with each other for different phase arrangements.

Magnetic flux lines around the overhead line are illustrated in Fig. 8.21 for the two phase conductor arrangements of symmetrical and optimum. Similar to the electric field intensities, magnetic flux densities caused by the optimum phase arrangement are lower than the symmetrical one which is shown in Fig. 8.22.

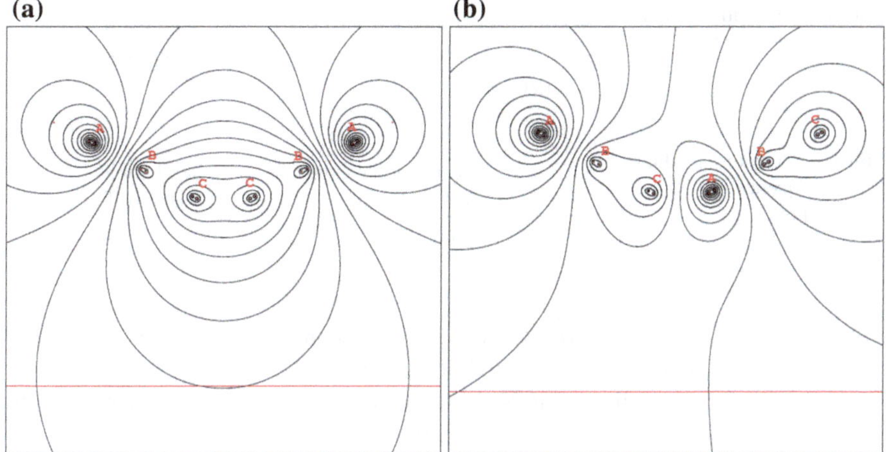

Fig. 8.21 Magnetic flux lines based on phase arrangements of **a** symmetrical and **b** optimum

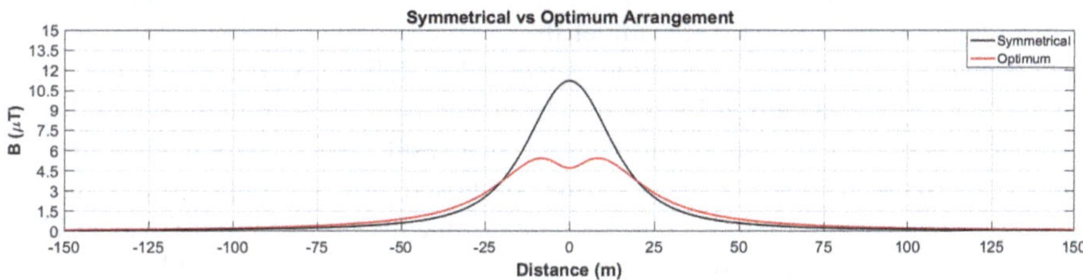

Fig. 8.22 Magnetic flux densities at one meter above the earth surface based on phase arrangements of symmetrical and optimum

8.6.3 Determination of Right-of-Way (ROW) Width

Right of way (ROW) determination can be affected by electrical and mechanical factors. In overall, electric and magnetic field limits should be taken into account the conductor swing factor. Since the fully composite pylon does not have suspension parts and conductors are fixed on the pylon by utilizing special conductor clamps, therefore the effects of conductors' swing can be ignored.

The required ROW for the overhead line composed of fully composite pylons can be determined by adopting acceptable limits (presented in ICNIRP standard) and comparing with the levels of electric and magnetic fields at one meter above the earth surface. The precaution values for exposure in the electric field are given in Table 8.7 which are based on the guidelines of ICNIRP standard [32].

According to Fig. 8.19, the electric field magnitudes under the overhead line is less than 5 kV_{rms}/m which is specified in Table 8.7. Therefore, the electric field

Table 8.7 Precaution values for low-frequency (50 Hz) electric field

Source	Occupational exposure (kV_{rms}/m)	General public exposure (kV_{rms}/m)
ICNIRP	10	5

Table 8.8 Precaution values for low-frequency (50 Hz) magnetic field

Source	Occupational exposure (μT)	General public exposure (μT)
ICNIRP	1000	200

magnitudes (at one meter above the earth surface) do not have any risk for occupational exposure and are also safe for public health.

In 2010, the new guidelines of ICNIRP reported new reference values for magnetic flux density which are given in Table 8.8 [36].

Generally, magnetic flux density in the vicinity and even directly below the overhead lines (at one meter above earth surface) is much lower than the precaution exposure limits [30]. Accordingly, Fig. 8.22 indicates that magnetic flux densities (at one meter above earth surface) under the overhead line are below 12 μT and are much lower than the precaution values specified in Table 8.8. Therefore, the emission of magnetic field caused by the overhead line is in the safe margin and do not have harmful effects for the public health. As a result, the low-frequency electromagnetic performance of the overhead line composed of fully composite pylons is in an acceptable level and there is no need to determine necessary right-of-way (ROW) width for the overhead line due to generated 50 Hz electromagnetic fields around the line.

8.7 Summary

In this chapter, the surface gradients on the phase conductors were computed by using finite element method. The obtained results showed that maximum electric field magnitudes on phase conductors are within the practical ranges. Based on the surface gradients, the A-weighted audible noise levels around the line were calculated by two methods of EPRI and BPA. The audible noises, which are caused to public annoyance, are occurred during foul and fair weathers. The generated audible noise at fair weather was much less than threshold value whereas the audible noise at foul weather was slightly higher than the recommended value. However, its value is comparable with regards to other traditional pylons.

Radio noise levels during different weather conditions were calculated based on two radio interference predicting methods: empirical and semi-analytical methods. The predicted radio noises by EdF and CIGRE give higher values in comparison to other methods whereas the noise levels calculated by IREQ and EPRI have lower values. Based on the threshold value of radio noise level, which is specified for the fair weather category, approximately all radio noise prediction methods estimate an acceptable radio interference performance for the fully composite pylon. Moreover, the calculated corona losses by EdF and BPA methods for fair weather category showed that corona loss of fully composite pylon is also within the recommended range.

Electric and magnetic fields around the line were calculated based on analytical and finite element methods. Two different phase conductor arrangements of symmetrical and optimum were assumed separately for the fully composite pylon. The results of finite element modeling coincided with the analytical ones, and magnetic field lateral profiles at both sides of the line were much lower than the precaution values for both phase conductor arrangements. Electric field lateral profiles were also below the constraint value for general public exposure. As a result, there is no need to define necessary right-of-way width at both sides of an overhead line composed of fully composite pylons for 50 Hz electromagnetic emissions.

References

1. R.G. Urban, *Power Line Corona Noise Prediction from Small Cage Measurement* (University of Stellebosch, 2004)
2. Q. Li, R. Shuttleworth, G. Zhang, S.M. Rowland, R.S. Morris, On calculating surface potential gradient of overhead line conductors, in *The 2012 IEEE International Symposium on Electrical Insulation (ISEI)* (2012)
3. Q. Li, *Acoustic Noise Emitted from Overhead Line Conductors* (The University Of Manchester, 2013)
4. Q. Li, S.M. Rowland, R. Shuttleworth, Calculating the surface potential gradient of overhead line conductors. IEEE Trans. Power Delivery **30**(1), 43–52 (2015)
5. D. Huang, J. Ruan, F. Huo, Study on the electromagnetic environment of 1000 kV AC double-circuit transmission lines in China, in *Power Systems Conference and Exposition, PSCE 2009* (IEEE/PES, 2009)
6. S. Hedtke, M.D. Pfeiffer, C.M. Franck, J. Bell, L. Zaffanella, J. Chan, Audible noise of hybrid AC/DC overhead lines: comparison of different prediction methods and conductor arrangements, in *Electric Power Research Institute (EPRI) HVDC & Flexible AC Transmission System (FACTS) Conference 2015* (2015)
7. C.L. Bak, S.D. Mikkelsen, C. Jensen, Overhead line audible noise measurements and calculation model for snow and frosty mist, in *International Symposium on High Voltage Engineering* (2005)
8. T. Jahangiri, *Electrical Design of a New, Innovative Overhead Line Transmission Tower Made in Composite Materials* (Aalborg University, 2018)
9. R. Lings, *EPRI AC Transmission Line Reference Book—200 kV and Above*, 3rd edn. (2005)
10. IEEE Std 656-1992: IEEE Standard for the Measurement of Audible Noise From Overhead Transmission Lines (1992)
11. J. Lundkvist, I. Gutman, L. Weimers, Feasibility study for converting 380 kV AC lines to hybrid AC/DC lines, in *EPRI's High-Voltage Direct Current & Flexible AC Transmission Systems Conference* (2009)
12. K.O. Papailiou, *CIGRE Green Book: Overhead Lines* (Springer, 2017)
13. F. Kiessling, P. Nefzger, J.F. Nolasco, U. Kaintzyk, *Overhead Power Lines, Planning, Design, Construction* (Springer, Heidelberg, 2003)
14. H.A. Roets, *Effect of Altitude on Audible Noise Generated by AC Conductor Corona* (Stellenbosch University, 2012)

15. D.E. Perry, An analysis of transmission, line audible noise levels based upon field and three-phase test line measurements. IEEE Trans. Power Appar.Us Syst. **PAS-91**(3), 857–865 (1972)
16. *Transmission Line Reference Book, 345 kV and Above*, 2nd edn. (EPRI, 1982)
17. CISPR/TR 18-2: Radio interference characteristics of overhead power lines and high-voltage equipment Part 2: methods of measurement and procedure for determining limits (2010)
18. CISPR/TR 18-1: Radio interference characteristics of overhead power lines and high-volt age equipment Part 1: description of phenomena (2010)
19. *Manitoba-Minnesota Transmission Project* (Exponent Inc., 2015)
20. A.Z. EL Dein, Prediction of egyptian 500 kV overhead transmission lines radio-interference by using the excitation function. WSEAS Trans. Power Syst. **9**, 479–485 (2014)
21. C. Zachariades, *Development of an Insulating Cross-arm for Overhead Lines*, (The University of Manchester, 2014)
22. S.N. Singh, *Electric Power Generation: Transmission and Distribution* (Prentice Hall India Pvt., 2004)
23. A. Chakrabarti, S. Halder, *Power System Analysis: Operation And Control*, 3rd edn. (PHI Learning Pvt. Ltd., 2010)
24. R. Sirjani, B. Hassanpour, Technical and economic assessment of upgrading a double-circuit 63 kV to a single-circuit 230 kV transmission line in Iran. Aust. J. Basic Appl. Sci. **5**(12), 2090–2097 (2011)
25. G. Filippopoulos, D. Tsanakas, G. Kouvarakis, J. Voyatzakis, M. Ammann, K. O. Papailiou, Optimum conductor arrangement of compact lines for electric and magnetic field minimization—calculations and measurements, in *Med Power* (2002)
26. K. Deželak, G. Štumberger, F. Jakl, Arrangements of overhead power line conductors related to the electromagnetic field limits, in *2010 Proceedings of the International Symposium Modern Electric Power Systems MEPS* (2010)
27. E.I. Mimos, D.K. Tsanakas, A.E. Tzinevrakis, Solutions for high voltage transmission in suburban regions regarding the electric and magnetic fields, in *World Automation Congress WAC 2008* (2008)
28. K. Deželak, G. Štumberger, F. Jakl, Reduction of electric and magnetic field emissions caused by overhead power lines, in *International Conference on Renewable Energies and Power Quality ICREPQ 2010* (2010)
29. W. Friedl, E. Schmautzer, G. Rechberger, A. Gaun, Constructional magnetic field reducing measures of high voltage overhead transmission lines, in *19th International Conference on Electricity Distribution CIRED 2007* (2007)
30. D. Reichelt, R. Scherer, R. Braunlich, T. Aschwanden, Magnetic field reduction measures for transmission lines considering power flow conditions, in *Transmission and Distribution Conference* (1996)
31. J.C. Salari, A. Mpalantinos, J.I. Silva, Comparative analysis of 2- and 3-D methods for computing electric and magnetic fields generated by overhead transmission lines. IEEE Trans. Power Delivery **24**(1), 338–344 (2009)
32. T. Jahangiri, M. Zavvari, V. Masouminia, Effects of catenary, compaction and phase conductor arrangement of double circuit lines on ROW width using calculations and FEM. Int. Rev. Electr. Eng. (IREE) **7**(4), 4980–4991 (2012)
33. R. Amiri, H. Hadi, M. Marich, The influence of sag in the electric field calculation around high voltage overhead transmission lines, in *IEEE Conference on Electrical Insulation and Dielectric Phenomena* (2006)
34. K. Deželak, G. Štumberger, F. Jakl, Emissions of electromagnetic fields caused by sagged overhead power lines, in *Przeglad Elektrotechniczny* (2011)
35. J.C. Fernandez, H.L. Soibelzon, Suggested location of the equivalent horizontal conductor when replacing catenary, in *18th International Conference and Exhibition on Electricity Distribution CIRED 2005* (2005)
36. E.I. Mimos, D.K. Tsanakas, A.E. Tzinevrakis, Electric and magnetic fields produced by 400 kV double circuit overhead lines—measurements and calculations in real lines and line models. CIGRE Sci. Eng. **5** (2016)

Chapter 9
Conclusion

9.1 Conclusions

Traditional overhead transmission line pylons, a key component of high voltage power systems, have a negative visual appearance on landscapes. For this reason, a fully composite pylon design has been introduced which has a better visual appearance with respect to traditional ones. The fully composite pylon design is a novel concept and its electrical and mechanical performance should be comprehensively evaluated. In this book, the electrical performance of fully composite pylon was investigated by using analytical and finite element methods and high voltage experimental method.

The major challenges in the design process of fully composite pylon were addressed in this book and resolved. One of the challenges was the electrical dimensioning of the fully composite pylon for which electrical clearances on the pylon should be determined. The electrical air clearances on the pylon must be enough to withstand against power frequency, switching and lightning overvoltages. Therefore, the air clearances on the pylon were calculated on the basis of an extensive insulation coordination study, and subsequently, internal and external clearance of the pylon at the tower top and mid-span were specified.

Another challenge was the selection of the material of the composite cross-arm core. Two kinds of candidate fibre reinforced plastic (FRP) materials were tested electrically and their important properties, including internal partial discharge (PD) performance, dielectric permittivity and dissipation factor were evaluated. Based on the test results, E-glass reinforced epoxy was recommended as the core material of the composite cross-arm. Three manufacturing methods for fibre reinforced plastic (FRP) materials were also compared and the filament winding method was recommended to the material supplier. An electrical-mechanical combined setup was designed and combined tests were performed on fibre reinforced plastic (FRP) materials and the effects of mechanical loading on their electrical behaviours were investigated.

Another challenge was the proper design of electrical insulations (phase-to-phase and phase-to-ground insulations) for the unibody cross-arm of fully composite pylon.

T. Jahangiri et al., *Electrical Design of a 400 kV Composite Tower*, Lecture Notes in Electrical Engineering 557,
https://doi.org/10.1007/978-3-030-17843-7_9

In this regard, the application of different shed profiles were analysed and a small and large shed profile were designed to the shed housing on the unibody cross-arm. One of the major concerns in the design of fully composite pylon was the electric field performance of the pylon. Numerous finite element analyses of the pylon were carried out to evaluate electric field and potential distribution around and inside the unibody cross-arm. Two design scenarios of utilizing internal or external ground connection for the shield wires were examined and it was recommended that an external ground connection should be chosen to provide ground potential access for shield wires. Different conductor clamp designs were considered for the attachment points between phase conductors and the unibody cross-arm and it was recommended that the utilization of fully steel conductor clamps as a feasible solution for the attachment points on the cross-arm. Moreover, electric field magnitudes in the different regions of interest on the fully composite pylon were calculated and compared with the threshold values and it was concluded that two corona rings with appropriate position and diameter should considered at the both sides of conductor clamps. In this regards, a multi-objective finite element analysis were done and the optimum positions and diameter of corona rings were found. As a result, it was observed that the fully composite pylon with optimized corona rings represents acceptable electric field performance.

In order to experimentally assessment of electric field magnitudes on the unibody cross-arm, a water-induced corona discharge setup was designed, with which a corona test was performed on an equivalent full-scale composite cross-arm segment (without corona rings) in wet conditions. Artificial rain was applied to the cross-arm segment with an equivalent phase-to-ground voltage. During this process, corona discharges from the cross-arm segment were measured and observed. With a stringent corona inception criterion of 10 pC, it is found out that corona is triggered from the cross-arm surface under nominal voltage in wet conditions. Thus, it was concluded that the design of the composite cross-arm (without corona rings) is not satisfactory and design improvement must be done to suppress the electric field magnitudes to an acceptable level. Therefore, based on the experimental results, it was verified that utilization of corona rings on both sides of conductor clamps are necessary to suppress electric field intensities on the cross-arm of fully composite pylon. Additionally, the suitability of the initially designed unibody cross-arm inclined angle from the ground plane was investigated. The water-induced corona discharge test and scale model test were performed with different cross-arm inclined angles. Based on test results, the initial inclined angle 30° is believed to be satisfactory if the electric field on the cross-arm surface can be restrained to an acceptable level.

The next challenge was the lightning shielding performance of fully composite pylon, which was calculated and evaluated by different methods. The conventional method of the electro-geometric model (EGM) was modified to be used for the assessment of effectiveness of lightning shielding system for the fully composite pylon. The feasibility of assigned negative shielding angle of −30° for the pylon was investigated and the outage rate regarding its application was calculated. The protected and unprotected zones on the fully composite pylon against vertical and possible horizontal lightning strikes were visualized for different lightning stroke currents. As a result, it was concluded that the fully composite pylon has acceptable lightning shielding performance with respect to other traditional pylons.

In order to experimental evaluation of the lightning protection performance of fully composite pylon, a scale model test was performed. It was experimentally proved that a shielding failure region exists at the center of the pylon and the experimental shielding failure rate for an overhead line composed of fully composite pylons is similar to that predicted by theoretical modified EGM method. Based on the scale model test, it was concluded that the shield wires arrangement on the fully composite pylon is satisfactory.

Finally, the environmental aspects of fully composite pylon were calculated based on analytical, empirical and finite element analysis methods. These aspects include radio interference, audible noise, corona loss and emission of electromagnetic field at the right-of-way of an overhead line composed of fully composite pylons. The calculated values were compared with the recommended threshold values and it was concluded that environmental effects of fully composite pylon are within acceptable ranges except audible noise level of the line which is slightly higher. The higher audible noise level of fully composite pylon is mainly due to its lower height. However, audible noise level of fully composite pylon is comparable with other traditional towers at the same voltage level.

In overall, the main challenges and relevant solutions in the theoretical design and experimental test processes of fully composite pylon were comprehensively presented and discussed in the content of this book which is structured as follows:

In Chap. 1, the scope of book including state of the art review on composite-based transmission pylons, introduction of Power Pylons of the Future (PoPyFu) project and fully composite pylon concept and research objectives of the research were presented.

In Chap. 2, Two kinds of fibre reinforced plastic samples—glass/epoxy and glass/vinlyester—manufactured by three different moulding methods—hand lay-up, vacuum assisted resin transfer and filament winding—were tested for internal partial discharge (PD) investigation. Test results indicated that the glass/epoxy sample manufactured by filament winding had highest partial discharge (PD) inception electric field for the same thickness. Its partial discharge (PD) inception electric field magnitude, 3.932 kV_{RMS}/mm for 1 mm samples and 3.126 kV_{RMS}/mm for 10 mm samples, are higher than the maximum electric field magnitude 0.6 kV_{RMS} in the fibre reinforced plastic (FRP) core of the composite cross-arm in service and also higher than the maximum allowable value 3 kV_{RMS}/mm. Meanwhile, circuits for dissipation factor and relative permittivity measurement were also implemented. The glass/epoxy sample manufactured by filament winding had the lowest dissipation factor −0.002 for 1 mm samples and 0.0025 for 10 mm samples—and highest relative permittivity −4.13 for 1 mm samples and 3.962 for 10 mm samples. Additionally, an electrical-mechanical combined test setup was established, aiming to investigate effects of mechanical loading on electrical behaviors of fibre reinforced plastic (FRP) composites. The combined test is a useful method to evaluate the electrical performance of the cross-arm core when exposed to electrical and mechanical stress simultaneously in service. In the experiments, only some immature combined tests results were presented, reflecting the effect of dynamic mechanical loading on partial discharge (PD) activities from glass/epoxy samples. Based on the electrical test results, glass/epoxy samples manufactured by filament winding method were selected as the most promising material for the fibre reinforced plastic (FRP) core of the composite cross-arm in the composite pylon, from electrical point of view.

In Chap. 3, the deterministic approach of insulation coordination study was briefly described. Statistical behavior and failure risk of insulation along with the effect of statistical withstand voltage and statistical overvoltage on the risk of failure were implied. The insulation coordination procedure to calculate required electrical clearances on the fully composite pylon was presented. To ensure and increase the reliability of long-term performance of the unibody cross-arm, reliable standard lightning impulse withstand voltage (LIWV) and switching impulse withstand voltage (SIWV) were adopted for the cross-arm's insulation. The most relevant gap configuration (conductor-conductor) was considered for the fully composite pylon to calculate accurate air clearances on the pylon. The calculated values were compared with the recommended values of standards and consequently, a conservative value and a calculated value were adopted for the phase-to-earth and phase-to-phase air clearances, respectively. Subsequently, internal and external air clearances at the tower top and within the mid-span were specified. According to the determined air clearances, basic dimensions of fully composite pylon were derived and delivered to the mechanical section for pursuing further researches in mechanical point of view.

In Chap. 4, the design of insulation for the unibody cross-arm of fully composite pylon was presented in more details. The design includes the calculation of creepage distances along the unibody cross-arm and allocating proper shed profiles on the shed housing of insulation. The required creepage distances on the unibody cross-arm were calculated for different pollution levels. For a specific site pollution severity (medium), an alternating large and small sheds was proposed for the insulation of shed housing of unibody cross-arm. The effectiveness and suitability of the shed profiles were evaluated based on relevant standards. Moreover, the electric field performance of shed profiles was investigated using finite element modelling of fully composite pylon.

Electric field criteria regarding the design of high voltage composite insulators were mentioned. Two design scenarios of utilizing internal and external ground connection for the shield wires were examined by performing numerous finite element analyses to evaluate electric field and potential distribution around and inside the unibody cross-arm. The feasibility of two design scenarios were discussed and it was concluded that the utilization of internal ground cable inside the unibody cross-arm imposes higher field intensities at air inside the cross-arm and can cause air insulation breakdown and puncture of the composite cross-arm under the energized parts. As a result, the external ground connection was proposed to be used for providing the ground potential for the shield wires.

Different conductor clamp designs with various material in used were considered for the attachment points between phase conductors and the unibody cross-arm. A fully steel conductor clamp concept together with corona rings was chosen as a feasible and suitable solution for the connection. Computation of electric field magnitudes in the different regions of interest on the unibody cross-arm showed that two corona rings should be installed at both sides of conductor clamps. In this regard, an optimization process was done to find appropriate dimension and positions for the corona rings. The outcome of the optimization was a design point in which fully composite pylon represents an acceptable electric field performance in all different regions of interests.

In Chap. 5, Corona discharges from the composite cross-arm was experimentally investigated in wet conditions and the electric field distribution around it was evaluated. An artificial rain was applied to a full-scale composite cross-arm segment (without corona rings), which represented for the phase-to-phase section in the composite pylon. The corona discharge from the cross-arm segment with a phase-to-ground voltage 420 kV_{RMS} was observed through a corona camera and measured by partial discharge (PD) measurement instrument. Test results showed that corona discharge was triggered from the composite cross-arm surface with water droplets on it at 379 kV_{RMS}. The maximum corresponding electric field at 0.5 mm from the weather sheath was computed as 0.256 kV/mm when corona was triggered. The experimental result indicated that corona discharge would be triggered from the composite cross-arm with the initial design at operation system voltage 420 kV_{RMS}. Based on test results, the conclusion can be drawn that the initial design of the composite cross-arm is not satisfactory. Two advices were given to improve the design of the composite cross-arm: (1) Corona rings must be applied to restrain the electric field magnitude on the cross-arm surface; (2) The conductor clamp profile should be considered more carefully in order to control the electric field magnitude to an acceptable level.

In Chap. 6, the basic concept of lightning shielding analysis of overhead transmission line using conventional electro-geometric (EGM) method was presented for the vertical lightning strikes. Since the conventional EGM method was not applicable for the evaluation of lightning performance of fully composite pylon, the conventional EGM was modified to be used for the assessment of lightning shielding performance of the pylon. The assigned shielding angle of $-60°$ for the fully composite pylon causes that an unprotected zone appears at the center of the pylon (against vertical lightning strikes). Therefore, instead of maximum shielding failure current in conventional EGM, minimum intersection current was used in the modified EGM method to compute shielding failure rate (SFR) and shielding failure flashover rate (SFFOR) of the pylon. The strengths and weakness of preliminary assigned shielding angle for the fully composite pylon were investigated. The obtained results for the shielding failure rate (SFR) of pylon showed that shielding failure rate of fully composite pylon is within acceptable range. Moreover, shielding failure flashover rate (SFFOR) of the pylon is theoretically zero. It proved that fully composite pylon provides reliable lightning shielding performance against vertical lightning strikes. The rolling sphere method was also implemented for the fully composite pylon to visualize the protection of phase conductors against any possible horizontal lightning strikes. Analytical results of modified EGM method and rolling sphere method (RSM) showed that fully composite pylon has acceptable lightning shielding performance against vertical and horizontal lightning strikes.

In Chap. 7, a scale model test was performed to experimentally investigate the shielding failure performance of the fully composite pylon and evaluate the electro-geometric model (EGM) results. Equivalent lightning flashes were applied to scale pylon model and scale overhead line. It was experimentally proved that the shielding failure region in the composite pylon existed in the pylon center. Meanwhile the number of lightning strikes to equivalent phase conductors were transformed into shielding failure rate and gap lengths in the pylon model were transformed into striking distances and then lightning current. As a result, the maximum shielding failure current was obtained based on the scale model test −4.45 kA with a fast-front impulse and 3.60 kA with a slow-front impulse. These values were approximately consistent with the value obtained by modified EGM. The scale model test results indicate that the lightning shielding failure performance of an overhead line composed of fully composite pylon is acceptable. Thus, the conclusion can be drawn that the shield wires arrangement on the fully composite pylon is satisfactory.

In Chap. 8, the environmental effects of fully composite pylon including audible noise, radio interference, corona losses and electromagnetic emissions were investigated. Based on the surface gradients on phase conductors, audible noise levels around the overhead line were calculated and the lateral profiles of audible noises were derived for different weather conditions. The audible noises, which are a cause to public annoyance, are occurred during foul and fair weathers. The generated audible noise at fair weather was much less than threshold value whereas the audible noise at foul weather was slightly higher than the recommended value. However, its value is comparable with regards to other traditional pylons. Radio noise levels during different weather conditions were calculated based on two radio interference predicting methods: empirical and semi-analytical methods. Lateral profiles of radio noise were presented and compared with recommended standard values. Based on the threshold value of radio noise level, which is specified for the fair weather category, approximately all radio noise prediction methods estimate an acceptable radio interference performance for the fully composite pylon. Corona loss of the line was calculated for fair weather condition. The results showed that corona loss of fully composite pylon is within the recommended range. Finally, electric and magnetic fields around the line were also computed by analytical and finite element approaches. Two different phase conductor arrangements of symmetrical and optimum were assumed separately for the fully composite pylon. The results of finite element modeling were coincided with the analytical ones, and magnetic field lateral profiles at both sides of the line were much lower than the precaution values for both phase conductor arrangements. Electric field lateral profiles were also below the constraint value for general public exposure. As a result, there is no need to define necessary right-of-way width at both sides of an overhead line composed of fully composite pylons for 50 Hz electromagnetic emissions.

In this chapter, the conclusions of book and future works were highlighted.

9.2 Future Challenges

Power Pylons of the Future (PoPyFu) project is an ongoing research project between Aalborg University, Technical University of Denmark, Bystrup Company and TUCO group. This project is expected to provide a futuristic competitive alternative for steel lattice towers, by presenting a production mature prototype as the basis for commercialization that has benefits to transmission system operators (TSOs) and the public especially in terms of visual appearance. The electrical performance of proposed fully composite pylon in PoPyFu project was investigated in this book, in which, the basic design concept along with some modification in the pylon design were considered and evaluated. However, the fully composite pylon is still in design and experimental test processes and its configuration, junctions, dimensions, and materials frequently change. These changes are due to the updates, which come from the development of pylon by electrical and mechanical partners. Therefore, the following researches should be done to consider and cover the new updates of design.

- The specified phase-to-earth and phase-to-phase air clearances are obtained from theoretical studies and are only based on the minimum required values. These values must be validated through full-scale experimental tests at high voltage laboratories. Moreover, the requirements for live line working and human safety should be taken into account.
- Any changes in the length of unibody cross-arm will lead to a change in the creepage distances, and subsequently, variations in the parameters of shed profiles on the shed housing. Conversely, in the case of utilizing other site pollution severities rather than medium level, the creepage distances and shed parameters will be changed and in some cases may impose higher air clearances between the conductors.
- Material properties of pylon's components have an influence on the electric field and potential distribution around and inside the fully composite pylon, therefore any changes in the material properties of unibody cross-arm and I-pole body of the pylon should be re-evaluated by using finite element method.
- The connection between shield wire's holder and the cross-arm has not been presented in more details. Thus, any update in this joint should be assessed by finite element method. In addition, a corona ring may also be needed to be installed at the updated joint.
- The connections between unibody cross-arm and I-pole body of the fully composite pylon has not been decided and finalized yet. Therefore, in the case of utilizing metal flanges for this purpose, additional corona ring may be required to modify electric field magnitudes around the joints.
- It is also assumed that phase conductors on the pylon are in the form of duplex bundles. In the case of utilizing triplex bundles for the phase conductors, electric field performance of the pylon should be re-evaluated by finite element method, especially around the steel conductor clamps.
- The cross-arm core is stressed by static and dynamic mechanical loads, due to the dead weight of conductors adding wind load and glaciation. Effects of dynamic

mechanical stress on the electrical performance of the full-scale cross-arm under mechanical stress is of importance. The filling of the hollow composite cross-arm has not been considered. Therefore, the most suitable filling materials and method must be developed for the composite cross-arm.

- The means of providing ground potential to shield wires have not been decided. As an important design development, a proper ground down lead must be determined in the future research. Its effectiveness and the visual impact to the whole tower configuration should be traded off. The surge impedance of the ground down lead also needs investigation to lower the back-flashover rate as much as possible.

Appendix A

See Tables A.1, A.2 and A.3.

Table A.1 Appendix: PD inception E-field E_{inc}, dissipation factor tan δ and relative permittivity ε_r of samples manufactured by hand lay-up molding method

Sample no.	Materials	E_{inc} [$kV_{rms/mm}$]	tan δ	εr
1	Glass/epoxy, 10 mm	0.62	0.0061	2.60
2		0.54	0.0064	2.67
3		0.60	0.0059	2.60
4		0.63	0.0059	2.67
5		0.58	0.0060	2.68
Average value	–	0.594	0.0061	2.64
Standard deviation		0.035	0.0002	0.04
1	Glass/epoxy, 1 mm	2.40	0.0032	3.17
2		2.00	0.0043	2.67
3		2.00	0.0052	3.09
4		2.40	0.0043	3.21
5		2.40	0.0052	2.94
Average value	–	2.24	0.0044	3.02
Standard deviation		0.22	0.0008	0.30
1	Glass/vinyl ester, 10 mm	0.60	0.0066	2.63
2		0.62	0.0074	2.80
3		0.61	0.0069	2.73
4		0.65	0.0073	2.76
5		0.5	0.0073	2.76
Average value	–	0.60	0.0071	2.74
Standard deviation		0.057	0.0003	0.06

(continued)

T. Jahangiri et al., *Electrical Design of a 400 kV Composite Tower*, Lecture Notes in Electrical Engineering 557,
https://doi.org/10.1007/978-3-030-17843-7

Table A.1 (continued)

Sample no.	Materials	E_{inc} [$kV_{rms/mm}$]	tan δ	εr
1	Glass/vinyl ester, 1 mm	1.10	0.0060	3.28
2		1.40	0.0062	3.42
3		1.20	0.0061	3.46
4		1.60	0.0057	3.50
5		1.00	0.0060	3.08
Average value	–	1.26	0.0060	3.35
Standard deviation		0.24	0.0002	0.17

Table A.2 Appendix: PD inception E-field E_{inc}, dissipation factor tan δ and relative permittivity ε_r of samples manufactured by vacuum assisted resin transfer molding method

No.	Materials	E_{inc} [kV_{rms}/mm]	tan δ	εr
1	Glass/epoxy, 10 mm	2.68	0.0025	3.56
2		3.3	0.0026	3.85
3		2.56	0.0028	3.66
4		2.85	0.0029	3.95
5		2.95	0.0025	4.10
Average value	–	2.87	0.0027	3.82
Standard deviation		0.28	0.0002	0.22
1	Glass/epoxy, 1 mm	3.20	0.0019	4.00
2		2.85	0.0022	3.56
3		3.10	0.0026	3.90
4		3.20	0.0028	4.20
5		3.50	0.0029	3.89
Average value	–	3.17	0.0025	3.91
Standard deviation		0.23	0.0004	0.23
1	Glass/vinyl ester, 10 mm	2.89	0.0026	3.88
2		3.67	0.0025	3.56
3		3.20	0.0024	3.89
4		1.90	0.0028	4.02
5		2.50	0.0022	4.20
Average value	–	2.83	0.0025	3.91
Standard deviation		0.67	0.0002	0.23
1	Glass/vinyl ester, 1 mm	3.81	0.0029	3.96
2		3.50	0.0020	3.56
3		2.90	0.0022	4.22
4		3.20	0.0022	4.01
5		3.20	0.0025	3.75
Average value	–	3.34	0.0024	3.90
Standard deviation		0.34	0.0004	0.25

Table A.3 Appendix: PD inception E-field E_{inc}, dissipation factor tan δ and relative permittivity ε_r of samples manufactured by filament winding molding method

No.	Materials	E_{inc} [kV_{rms}/mm]	tan δ	εr
1	Glass/epoxy, 10 mm	2.71	0.0026	3.90
2		3.02	0.0028	4.00
3		2.80	0.0024	4.10
4		3.65	0.0021	3.85
5		3.45	0.0026	3.96
Average value	–	3.13	0.0025	3.96
Standard deviation		0.41	0.0003	0.10
1	Glass/epoxy, 1 mm	3.46	0.0021	4.12
2		3.80	0.0022	4.23
3		4.70	0.0026	3.95
4		4.00	0.0020	4.12
5		3.70	0.0021	4.23
Average value	–	3.93	0.0022	4.13
Standard deviation		0.47	0.0002	0.11
1	Glass/vinyl ester, 10 mm	2.80	0.0023	4.02
2		2.85	0.0026	3.86
3		2.65	0.0025	4.12
4		2.30	0.0027	4.22
5		2.70	0.0029	4.25
Average value	–	2.66	0.0026	4.09
Standard deviation		0.22	0.0002	0.16
1	Glass/vinyl ester, 1 mm	3.15	0.0025	3.94
2		3.26	0.0024	4.02
3		2.90	0.0026	4.36
4		2.80	0.0022	4.25
5		3.02	0.0025	4.44
Average value	–	3.03	0.0024	4.20
Standard deviation		0.19	0.0002	0.22

Appendix B

See Figures B.1 and B.2 and Tables B.1 and B.2.

T. Jahangiri et al., *Electrical Design of a 400 kV Composite Tower*, Lecture Notes in Electrical Engineering 557,
https://doi.org/10.1007/978-3-030-17843-7

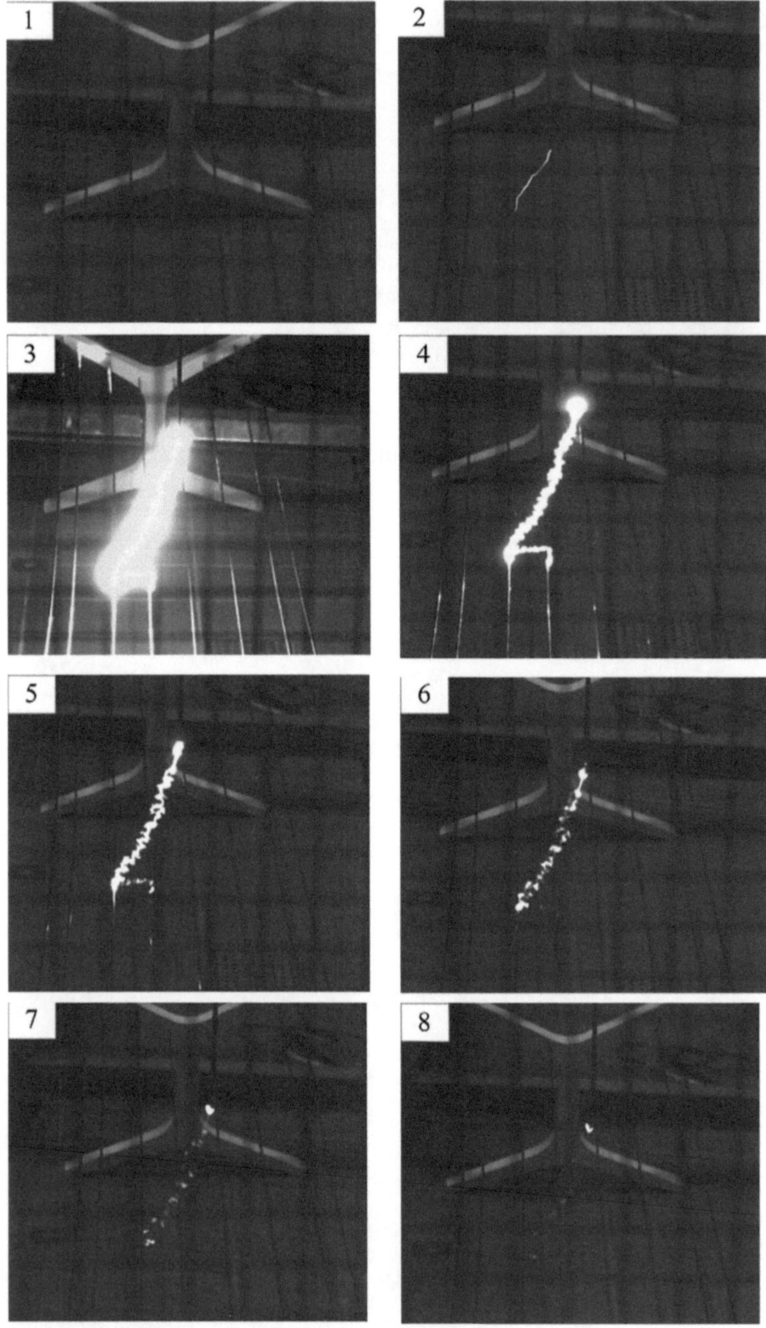

Fig. B.1 Appendix: A typical temporal development of a discharge in a chronological order (1–8) with the fast-front impulse in the scale model test (The shielding angle is −60°)

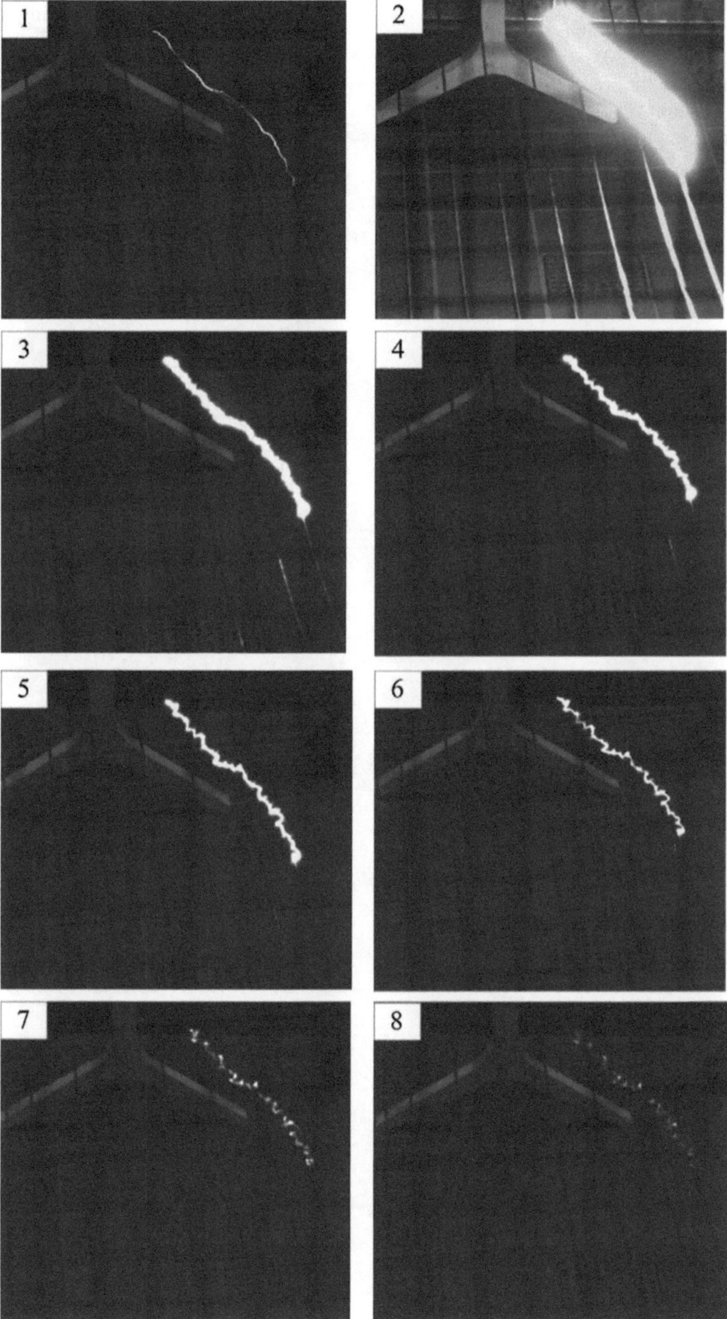

Fig. B.2 Appendix: A typical temporal development of a discharge in a chronological order (1–8) with the slow-front impulse in the scale model test (The shielding angle is $-60°$)

Table B.1 Appendix: Shielding failure rate with a shielding angle of −30° corresponding to the electrode location (x, y) with the fast-front impulse in the scale model test

y	x	Shielding failure rate SFR [%]
15	0	90
	3	78
	5	68
	7	26
	10	16
20	0	76
	3	58
	5	42
	7	12
	10	0
25	0	66
	3	48
	5	34
	7	10
	10	0
30	0	22
	3	18
	5	4
	7	0
	10	0
35	0	24
	3	12
	5	4
	7	0
	10	0
40	0	16
	3	4
	5	2
	7	0
	10	0
45	0	4
	3	2
	5	0
	7	0
	10	0

Table B.2 Appendix: Shielding failure rate with a shielding angle of −30° corresponding to the electrode location (x, y) with the slow-front impulse in the scale model test

y	x	Shielding failure rate SFR [%]
15	0	92
	3	22
	5	14
	7	8
	10	0
20	0	40
	3	14
	5	0
	7	0
	10	0
25	0	20
	3	8
	5	0
	7	0
	10	0
30	0	24
	3	4
	5	6
	7	0
	10	0
35	0	6
	3	0
	5	0
	7	0
	10	0
40	0	2
	3	0
	5	0
	7	0
	10	0
45	0	0
	3	0
	5	0
	7	0
	10	0